International Association of Fire Chiefs

International Society of Fire Service Instructors

Fire Protection Association

Fire Service Instructor
Principles and Practice

Student Workbook

JONES AND BARTLETT PUBLISHERS
Sudbury, Massachusetts
BOSTON   TORONTO   LONDON   SINGAPORE

World Headquarters
Jones and Bartlett Publishers
40 Tall Pine Drive
Sudbury, MA 01776
978-443-5000
info@jbpub.com
www.jbpub.com

National Fire Protection
Association
1 Batterymarch Park
Quincy, MA 02164-7471
www.nfpa.org

International Association of
Fire Chiefs
4025 Fair Ridge Drive
Fairfax, VA 22033
www.iafc.org

International Society of
Fire Service Instructors
2425 Highway 49 East
Pleasant View, TN 37146
www.isfsi.org

Jones and Bartlett Publishers Canada
6339 Ormindale Way
Mississauga, Ontario L5V 1J2
Canada

Jones and Bartlett Publishers International
Barb House, Barb Mews
London W6 7PA
United Kingdom

Jones and Bartlett's books and products are available through most bookstores and online booksellers. To contact Jones and Bartlett Publishers directly, call 800-832-0034, fax 978-443-8000, or visit our website www.jbpub.com.

> Substantial discounts on bulk quantities of Jones and Bartlett's publications are available to corporations, professional associations, and other qualified organizations. For details and specific discount information, contact the special sales department at Jones and Bartlett via the above contact information or send an email to specialsales@jbpub.com.

Copyright © 2009 by Jones and Bartlett Publishers, LLC and the National Fire Protection Association®.

All rights reserved. No part of the material protected by this copyright may be reproduced or utilized in any form, electronic or mechanical, including photocopying, recording, or by any information storage and retrieval system, without written permission from the copyright owner.

The procedures and protocols in this book are based on the most current recommendations of responsible sources. The International Association of Fire Chiefs (IAFC), National Fire Protection Association (NFPA®), International Society of Fire Service Instructors (ISFSI), and the publisher, however, make no guarantee as to, and assume no responsibility for, the correctness, sufficiency, or completeness of such information or recommendations. Other or additional safety measures may be required under particular circumstances.

Editorial Credits

Authors: Alan E. Joos and Forest F. Reeder

Production Credits

Chief Executive Officer: Clayton Jones
Chief Operating Officer: Don W. Jones, Jr.
President, Higher Education and Professional Publishing: Robert W. Holland, Jr.
V.P., Sales: William J. Kane
V.P., Design and Production: Anne Spencer
V.P., Manufacturing and Inventory Control: Therese Connell
Publisher: Kimberly Brophy
Senior Acquisitions Editor—Fire: William Larkin
Managing Editor: Carol Guerrero
Associate Editor: Karen Greene
Production Manager: Jenny L. Corriveau
Director of Marketing: Alisha Weisman
Marketing Manager: Brian Rooney
Composition: Publishers' Design and Production Services, Inc.
Cover Design: Kristin E. Ohlin
Cover Image: © Courtesy of the University of Nevada, Reno Fire Science Academy
Printing and Binding: Courier Stoughton
Cover Printing: Courier Stoughton

ISBN-13: 978-0-7637-6035-9

6048

Printed in the United States of America
12 11 10 09 08   10 9 8 7 6 5 4 3 2 1

Table of Contents

> Note to the student. Exercises indicated with a ⬛II are specific to the Fire Service Instructor II Level.

Part I: Introduction
Chapter 1: Today's Emergency Services Instructor — 1
Chapter 2: Legal Issues — 6

Part II: Instructional Delivery
Chapter 3: Methods of Instruction — 12
Chapter 4: The Learning Process — 18

Part III: Practical Applications
Chapter 5: Communication Skills for the Instructor — 24
Chapter 6: Lesson Plans — 30
Chapter 7: The Learning Environment — 36
Chapter 8: Training Today: Multimedia Applications — 42
Chapter 9: Safety During the Learning Process — 48

Part IV: Evaluation and Testing
Chapter 10: Evaluating the Learning Process — 54
Chapter 11: Evaluating the Fire Service Instructor — 60

Part V: Program Management
Chapter 12: Managing the Training Team — 66
Chapter 13: The Learning Process Never Stops — 72

Fire Service Instructor II Student Application Package — 79

Answer Key

Part I: Introduction
Chapter 1: Today's Emergency Services Instructor — 97
Chapter 2: Legal Issues — 98

Part II: Instructional Delivery
Chapter 3: Methods of Instruction — 100
Chapter 4: The Learning Process — 101

Part III: Practical Applications
Chapter 5: Communication Skills for the Instructor — 102
Chapter 6: Lesson Plans — 103
Chapter 7: The Learning Environment — 104
Chapter 8: Training Today: Multimedia Applications — 105
Chapter 9: Safety During the Learning Process — 106

Part IV: Evaluation and Testing
Chapter 10: Evaluating the Learning Process — 107
Chapter 11: Evaluating the Fire Service Instructor — 108

Part V: Program Management
Chapter 12: Managing the Training Team — 109
Chapter 13: The Learning Process Never Stops — 110

Additional Student Resources

To help students retain the most important information and assist them in preparing for exams, Jones and Bartlett Publishers has developed the following student resources.

www.fire.jbpub.com

This site has been specifically designed to complement *Fire Service Instructor: Principles and Practice* and is regularly updated. Resources available include:

- **Chapter Pretests** that prepare students for training. Each chapter has a pretest and provides instant results, feedback on incorrect answers, and page references for further study.
- **Interactivities** that allow students to reinforce their understanding of the most important concepts in each chapter.
- **24 sample Fire Service Instructor I Presentations** are included to help students practice their presentations skills.
- **Hot Term Explorer,** a virtual dictionary, allowing students to review key terms, test their knowledge of key terms through quizzes and flashcards, and complete crossword puzzles.

Exam Prep: Fire Instructor I & II, Second Edition

ISBN-13: 978-0-7637-5837-0

Exam Prep: Fire Instructor I & II, Second Edition is designed to prepare you to sit for a Fire Instructor I & II certification, promotion, or training examination by including the same type of multiple-choice questions you are likely to encounter.

The manual follows the Systematic Approach to Examination Preparation, developed by Performance Training Systems, Inc., to help improve examination scores. The practice examinations were written by fire and emergency service personnel and the content was validated through current reference materials and technical review committees. Your exam performance will improve after using this system.

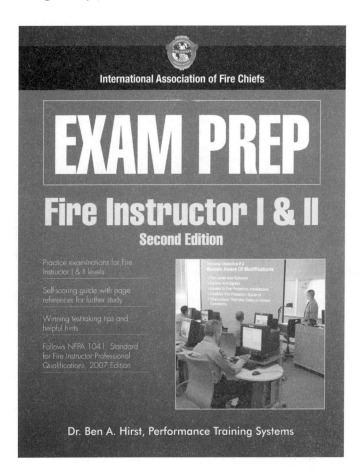

Workbook Activities

The following activities have been designed to help you. Your instructor may require you to complete some or all of these activities as a regular part of your instructor training program. You are encouraged to complete any activity that your instructor does not assign as a way to enhance your learning in the classroom.

Chapter Review

The following exercises provide an opportunity to refresh your knowledge of this chapter.

Multiple Choice

Read each item carefully, and then select the best response.

_____ 1. In past years, where did firefighter training typically take place?
   A. On the job
   B. At the fire academy
   C. In trade schools
   D. As individual study

_____ 2. Fire service instructors should have experience in which of the following?
   A. Public speaking
   B. The subject matter that they are teaching
   C. Application of material to job skills
   D. All of the above

_____ 3. Which instructor attribute brings excitement to the training environment?
   A. Apathy
   B. Monotone voice
   C. Motivation
   D. Lack of confidence

_____ 4. Which NFPA standard governs the duties, knowledge, and skills of the fire service instructor?
   A. 1001
   B. 1021
   C. 1041
   D. 1500

Chapter 1

Today's Emergency Services Instructor

_____ 5. The Fire Service Instructor I has job performance duties that include which of the following:
   A. To create lesson plans
   B. To write performance objectives
   C. To develop testing evaluation instruments
   D. To present instruction from a prepared lesson plan

_____ 6. What type of diagram is used by an organization to identify the various responsibilities and duties of an individual within a department or work area?
   A. Organizational chart
   B. Flow chart
   C. Venn diagram
   D. Standard operating procedure

_____ 7. Why are fire service instructors considered to be mentors of the fire service?
   A. Because of their position in the department's organizational chart
   B. Because of their job description duties and responsibilities
   C. Because they can observe the development of job skills from recruit to journeyman
   D. Because of the rank designation of the position of instructor

_____ 8. Which attribute identifies a proper learning environment for classroom and drill ground activities?
   A. Safe from potential hazards
   B. Distraction free
   C. Environmentally comfortable
   D. All of the above

_____ 9. What term is used to identify a process that is used to provide continuity of the organization and security for the community over time?
   A. Strategic planning
   B. Succession planning
   C. Candidate testing
   D. Mission statement

10. What type of documentation is issued after a person has completed acquisition of the required knowledge in a particular field?
   A. Degree
   B. Certification
   C. Certificate
   D. Award

11. What type of documentation is issued to a person who attends a learning event that has no testing or performance requirement?
   A. Degree
   B. Certification
   C. Certificate
   D. Award

12. What process will be used to help the instructor stay current on trends and techniques needed throughout his or her career?
   A. Continuing education
   B. Recordkeeping
   C. Certification
   D. Accountability

Matching

Match the following role or responsibilities with the correct level of instructor responsibility.

A. Instructor I
B. Instructor II
C. Both Instructor I and II

___ 1. Manage instructional resources, staff, facilities, and records and reports
___ 2. Develop student evaluation instruments
___ 3. Adjust presentations to students' different learning styles
___ 4. Administer and grade student evaluation instruments
___ 5. Conduct classes using a lesson plan
___ 6. Schedule instructional sessions
___ 7. Review and adapt prepared instructional materials
___ 8. Develop instructional material for specific topics

Match the following fire service instructor roles with the correct definition.

A. Leader
B. Mentor
C. Coach
D. Evaluator
E. Teacher

___ 9. Identifies future leaders of the organization by observing raw talent and ongoing growth
___ 10. Prepares crews and members for operations by building team skills through practice and repetition
___ 11. Sets the example for all firefighters to follow in terms of performance excellence
___ 12. Presents new skills and abilities in a variety of learning environments
___ 13. Identifies the level of proficiency that an individual possesses after learning has taken place and provides feedback on performance

True/False

If you believe the statement to be more true than false, write the letter *T* in the space provided. If you believe the statement to be more false than true, write the letter *F*.

_____ 1. A Fire Service Instructor I prepares lesson plans for use in fire service training.

_____ 2. A Fire Service Instructor II has many diverse roles and responsibilities in the management of instructional resources.

_____ 3. The Fire Service Instructor II must also perform all of the duties of the Instructor I.

_____ 4. The classroom and drill ground should provide a safe, comfortable, and distraction-free environment whenever possible.

_____ 5. There are many physical factors such as seating arrangement that can both positively and negatively impact the success of the learning environment.

_____ 6. Good instructors and students alike continually place themselves in learning situations.

_____ 7. Training programs are immune from national standards.

_____ 8. Ethics are not typically an issue that an instructor must deal with in instruction.

_____ 9. There are many issues of confidentiality that the instructor will have to deal with, including information regarding learning disabilities.

_____ 10. The fire service instructor's department or organizational rank does not give an element of credibility to the instructors.

Fill-in

Read each item carefully, and then complete the statement by filling in the missing word(s).

1. The _____ for instructors are documented in *NFPA 1041*.

2. As a fire service instructor, you must be aware that your _____ or _____ rests with the degree of effort and preparation put forth at the beginning of your career.

3. _____ set the bar for fire service instructors and are intended to maintain a high level of proficiency and knowledge.

4. _____ spell out what is acceptable and what is unacceptable within a fire department.

5. All training sessions must be _____, along with a factual listing of the objectives that were accomplished during the training session.

Short Answer

Complete this section with short written answers using the space provided.

1. Define the roles and responsibilities of the Fire Service Instructor I.

2. Define the roles and responsibilities of the Fire Service Instructor II.

3. Describe the physical elements of a classroom.

4. Identify and explain the five major roles of the fire service instructor.

5. Identify and explain four issues of ethics for fire service instructors.

Instructor Applications

Complete this section with short written answers using the space provided.

1. List the changes in training delivery that you have seen or experienced since you began your career in the fire service.

   _____
   _____
   _____
   _____
   _____

2. Select an industry Web site or reference materials, and identify at least three hot topics facing instructors today.

   _____
   _____
   _____
   _____
   _____

3. Identify a new piece of equipment purchased by your department, and describe how initial training was accomplished.

   _____
   _____
   _____
   _____
   _____

4. Ask the newest or youngest member of your organization to describe how he or she was educated at the last school that he or she attended. Identify similarities and differences between how fire service training and traditional education is presented.

   _____
   _____
   _____
   _____
   _____

Workbook Activities

The following activities have been designed to help you. Your instructor may require you to complete some or all of these activities as a regular part of your instructor training program. You are encouraged to complete any activity that your instructor does not assign as a way to enhance your learning in the classroom.

Chapter Review

The following exercises provide an opportunity to refresh your knowledge of this chapter.

Multiple Choice

Read each item carefully, and then select the best response.

_____ 1. Which of the following sources of law sets expectation for and restrictions on the fire service instructor's conduct?
   A. Individual fire department policy
   B. Federal law
   C. State law
   D. All of the above

_____ 2. According to the Americans with Disabilities Act, which of the following would be defined as an impairment that is covered by this act?
   A. Not being able to swim
   B. Inability to don an SCBA in 60 seconds after repeated training
   C. A hearing or speaking disorder
   D. Compulsive gambling

_____ 3. What is the term that is used to define what an employer does to assist the individual with a disability covered by the Americans with Disabilities Act?
   A. Provide reasonable accommodation
   B. Physical therapy
   C. Undue hardship
   D. Eligibility determination

_____ 4. What does Title VII prohibit against in the event that an individual files a charge of discrimination?
   A. Compensation
   B. Retaliation
   C. Discrimination
   D. Aggressive behavior

Chapter 2

Legal Issues

_____ 5. What is the simplest form of responsibility that also places an assignment of blame on an individual or an organization?
   A. Liability
   B. Malfeasance
   C. Misfeasance
   D. Tort

_____ 6. If an organization passed a rule that stated it was an "English-only" employer, it could be found guilty of what type of discrimination?
   A. Civil rights
   B. Age
   C. Religious
   D. National origin

_____ 7. Which type of wrong can be alleged in civil cases when an injury occurs that may be equated to the term accident?
   A. Negligence
   B. Malfeasance
   C. Responsibility
   D. Tort liability

_____ 8. One of the first lines of defense in the event that a lawsuit is filed is
   A. Proper recordkeeping
   B. Purchasing liability insurance
   C. Having a witness for everything you do
   D. Citing case studies when presenting information

_____ 9. The best practice for an organization is to put what type of policy in place to prevent discrimination?
   A. A sexual harassment policy
   B. An Americans with Disabilities Act compliance policy
   C. An antidiscrimination policy
   D. Not hiring protected class individuals

_____ 10. What types of files and records must be kept secure because they may be protected by state and federal privacy laws?
   A. Personnel files
   B. Hiring files
   C. Disciplinary files
   D. All of the above

Matching

Match the following types of law with the correct definition.

A. Standards
B. Statutes
C. Codes/regulations

_____ 1. Created by legislative action and can be either state or federal
_____ 2. Any rule, guideline, or practice recognized as being established by authority or using common language
_____ 3. Created by an administrative agency with the authority to do so

Match the following type of law with the example that best identifies it.

D. Standards
E. Statutes
F. Codes/regulation

_____ 4. Building code
_____ 5. NFPA 1403
_____ 6. Americans with Disabilities Act

Match the following Federal Employment Law with the brief summary of its content.

G. Age Discrimination in Employment Act
H. Equal Pay Act
I. Americans with Disabilities Act
J. Section 1983 of the Civil Rights Act of 1964
K. Title VII of the Civil Rights Act

_____ 7. Prohibits discrimination against a qualified person because of a disability
_____ 8. Prohibits discrimination based on race, gender, or national origin
_____ 9. Requires that all persons regardless of gender be paid at the same rate when performing the same job
_____ 10. Prohibits discrimination that creates a hostile work environment
_____ 11. Prohibits discrimination against persons older than the age of 40 years

Match the type of records or reports with the typical contents of those files.

L. Personnel files
M. Hiring files
N. Disciplinary files
O. Training files

_____ 12. Certification records and certificates of class completion
_____ 13. Reports of reprimands or punishment
_____ 14. Individual test scores, pre-employment physical exams
_____ 15. Birth certificate, social security number, and dependent information

True/False

If you believe the statement to be more true than false, write the letter *T* in the space provided. If you believe the statement to be more false than true, write the letter *F*.

_____ 1. Every fire department must have a policy in place prohibiting sexual harassment in the workplace.

_____ 2. In a sexual harassment situation, parties must be of the opposite gender.

_____ 3. Many materials you may want to use for training purposes are covered by copyright.

_____ 4. Fire service instructors who joke with students regarding religion or nationality cannot be liable for those comments because they are known to use comedy as part of their presentation style.

_____ 5. Training and education is part of an organization's risk management program.

_____ 6. Proper ethical conduct should be the goal of all those involved in the fire service.

_____ 7. Gross negligence is defined as an act that if not intentional shows utter indifference or conscious disregard for the safety of others.

_____ 8. Failing to follow a written protocol that creates a substantial risk to another person can be termed as willful and wanton conduct.

_____ 9. It is a best practice to seek written permission to use copies of written materials for classes.

_____ 10. It would be a violation of copyright law to purchase one copy of a textbook and to make copies of it for other students in the class.

Fill-in

Read each item carefully, and then complete the statement by filling in the missing word(s).

1. Race and color harassment includes _____, jokes, and derogatory or offensive comments.

2. The Americans with Disabilities Act defines _____ as a modification or adjustment to a job or work environment that will enable a qualified applicant or employee to perform the job.

3. Sexual harassment consists of _____, requests for sexual favors, and other verbal or physical contact of a sexual nature.

4. Simply defined, _____ consists of improper or wrongful performance of a lawful act without intent through mistake or carelessness.

5. Failure to abide by the _____ provisions required by law or your department could lead to a liability or adverse action against you.

Short Answer

Complete this section with short written answers using the space provided.

1. Describe how to document when an injury occurs to a participant or observer of a training session.

    _____
    _____
    _____
    _____

2. Describe the training records and reports required by the fire department.

    _____
    _____
    _____
    _____

3. Define the types of laws that apply to the fire service instructor.

    _____
    _____
    _____
    _____

4. List examples of the types of disabilities that are covered by the Americans with Disabilities Act.

    _____
    _____
    _____
    _____

5. List examples of the types of materials that are covered by copyright laws.

    _____
    _____
    _____
    _____

Instructor Applications

Discuss your response to these instructor applications to assist you in developing knowledge about your responsibilities as a fire service instructor.

1. Review your training policy for requirements in documentation of training sessions and recordkeeping.

2. List the local ordinances, laws, and standards that apply to the development, delivery, and recordkeeping of training.

3. Identify the written policies and procedures in place in your organization that address the laws and standards that are discussed in this chapter.

Workbook Activities

The following activities have been designed to help you. Your instructor may require you to complete some or all of these activities as a regular part of your instructor training program. You are encouraged to complete any activity that your instructor does not assign as a way to enhance your learning in the classroom.

Chapter Review

The following exercises provide an opportunity for you to refresh your knowledge of this chapter.

Multiple Choice

Read each item carefully, and then select the best response.

_____ 1. What factors combine to determine performance in training?
   A. Learning and motivation
   B. Cognitive and psychomotor
   C. Time and distance
   D. Interaction and participation

_____ 2. How many types of adult learners are identified in this chapter?
   A. 1
   B. 2
   C. 3
   D. 4

_____ 3. Which of the following is not a type of adult learner?
   A. Self-centered
   B. Goal-oriented
   C. Learner-oriented
   D. Activity-oriented

_____ 4. One of the primary questions that affects how you design your class is answered by asking which question?
   A. When will the class end?
   B. Why is the adult learner in the class?
   C. What will the student learn by attending?
   D. How does the classroom have to be arranged?

Chapter 3

Methods of Instruction

_____ 5. One attribute of an effective instructor that may be contagious in the classroom that ultimately improves the learning environment is called what?
   A. Learning style
   B. Empathy
   C. Voice and eye contact
   D. Motivation

_____ 6. Each adult learner has a distinct and preferred way of perceiving, organizing, and retaining experiences. This is called your what?
   A. Motivation
   B. Learning style
   C. Thought process
   D. Perception

_____ 7. Which generation of adult learners is characterized as the senior membership of most departments and usually offers many experiences and much knowledge?
   A. Generation X
   B. Generation Y
   C. Baby boomers
   D. Generation Z

_____ 8. Which generation of adult learners is characterized as one that is conditioned to expect immediate gratification, that is ambitious, and who tends to include independent problem solvers?
   A. Generation X
   B. Generation Y
   C. Baby boomers
   D. Generation Z

_____ 9. Which generation of adult learners is known for holding a wide range of opinions on a wide range of issues and may be considered to be pampered and self-centered?
   A. Generation X
   B. Generation Y
   C. Baby boomers
   D. Generation Z

_____ 10. What type of disruptive student is preoccupied with anything other than the lesson being presented?
   A. Monopolizer
   B. Expert
   C. Day dreamer
   D. Historian

_____ 11. What type of disruptive student may become confrontational in order to show off their brain power or perceived knowledge of a particular subject?
   A. Monopolizer
   B. Expert
   C. Day dreamer
   D. Historian

_____ 12. When dealing with disruptive behavior during instruction, one of the most important rules to remember is that
   A. The students needs come first.
   B. You are in charge of the classroom.
   C. Unproductive behavior must be addressed after the class is over.
   D. Everyone has a different learning style, and some people will simply act out.

Matching

Place the sequence of how effective adult learning takes place into the correct order.

A. 1
B. 2
C. 3
D. 4

_____ 1. Adult learners use the new content, practicing how it can be applied to real life.

_____ 2. Seeing that "newly" learned information has real-life connections prepares the adult learner for the next step—learning something new.

_____ 3. Adult learners take the material learned in the classroom and apply it to the real world, including in situations that were not always covered in the original lesson.

_____ 4. Adult learners begin with what they already know, feel, or need, based on the groundwork that was laid before what is currently taking place. In other words, learning does not take place in a vacuum.

Match the type of adult learner with the way that he or she learns new information.

E. Learner-oriented
F. Goal-oriented
G. Activity-oriented

_____ 5. Acquires knowledge with the goal of improving job skills or learning a new skill

_____ 6. Requires personal protective time with control over content and learning style

_____ 7. Likes to be involved in the learning

Match the percentage of what we remember to how we receive information.

H. 10 percent
I. 20 percent
J. 30 percent
K. 50 percent
L. 70 percent
M. 90 percent

_____ 8. What we see and hear
_____ 9. What we read
_____ 10. What we say while doing
_____ 11. What we hear
_____ 12. What we say
_____ 13. What we see

True/False

If you believe the statement to be more true than false, write the letter *T* in the space provided. If you believe the statement to be more false than true, write the letter *F*.

_____ 1. The type of learner you are affects the way learning takes place.
_____ 2. The monopolizer is a disruptive student who has some experience and wants to make sure that everyone knows it.
_____ 3. Behavioral problems displayed by disruptive students may be rooted in other problems.
_____ 4. Calling students by their name during instruction makes them feel important and may get them more involved in learning.
_____ 5. A generation X learner represents a change in learning methods that may involve the use of technology more than others.
_____ 6. Adult learners prefer to have goals set for them in their learning environment.
_____ 7. In the adult learning environment, the adult likes to begin with what they already know or to start from their comfort zone.
_____ 8. As a fire service instructor, you have the responsibility to ensure that the learning environment is appropriate for all of your students, even those not in the mainstream.
_____ 9. Adult learning does not need to be problem or experience centered to be effective.
_____ 10. We remember 90 percent of what we say while doing.
_____ 11. Andragogy is an attempt to identify the way in which adults learn.
_____ 12. All adult learners learn alike.

Fill-in

Read each item carefully, and then complete the statement by filling in the missing word(s).

1. Learning is the potential behavior, whereas _____ is the behavior activator.

2. The term _____ has been used to describe the generation consisting of those people whose teenaged years were touched by the 80s and were the first generation in which both parents typically worked.

3. A _____ can be used to signal an important point in your lecture.

4. We remember only _____ of what we read, but remember _____ of what we say while doing.

5. Factors such as participation in _____ since high school will affect the adult learner's ability to learn.

Short Answer

Complete this section with short written answers using the space provided.

1. Describe motivational techniques that affect the learning environment.

   _____
   _____
   _____
   _____

2. Describe how to adjust the classroom presentation to meet the needs of specific learning objectives.

   _____
   _____
   _____
   _____

3. Describe the principles of adult learning.

   _____
   _____
   _____
   _____

4. Describe how to deal with disruptive or unsafe behavior by students during instruction.

   _____
   _____
   _____
   _____

5. Discuss the characteristics of each generational learner.

   _____
   _____
   _____
   _____

Instructor Applications

Discuss your response to these instructor applications to assist you in developing your knowledge of your responsibilities as a fire service instructor.

1. Describe the most effective and least effective training sessions that you have attended, and discuss which methods of instruction were appropriate or inappropriate and that may have influenced your impression of the training session.

   _____
   _____
   _____
   _____
   _____
   _____
   _____

2. Review a list of upcoming training sessions scheduled for your department, and determine the best methods of instruction that should be used to achieve the best outcomes for them.

   _____
   _____
   _____
   _____
   _____
   _____
   _____

3. Rearrange your typical classroom seating arrangement to accommodate various methods of discussion such as a lecture, demonstration, or group discussion.

   _____
   _____
   _____
   _____
   _____
   _____
   _____

Workbook Activities

The following activities have been designed to help you. Your instructor may require you to complete some or all of these activities as a regular part of your instructor training program. You are encouraged to complete any activity that your instructor does not assign as a way to enhance your learning in the classroom.

Chapter Review

The following exercises provide an opportunity to refresh your knowledge of this chapter.

Multiple Choice

Read each item carefully, and then select the best response.

_____ 1. According to Edward Thorndike, how many laws of learning apply to how behavior is changed through instruction?
  A. 1
  B. 2
  C. 5
  D. 6

_____ 2. Which perspective on learning is reflected when an intellectual process takes place to improve the way you mentally view information?
  A. Psychomotor
  B. Behaviorist
  C. Cognitive
  D. Competency

_____ 3. Which perspective on learning is tied to skills or hands-on training?
  A. Psychomotor
  B. Behaviorist
  C. Cognitive
  D. Competency

_____ 4. According to the laws of learning, which of the following represents the law of recency?
  A. Learning accompanied with satisfaction
  B. The more recent the practice, the more effective the performance
  C. Applying real-life experiences in realistic settings
  D. A person can learn when physically and mentally ready to learn

Chapter 4

The Learning Process

_____ 5. According to the laws of learning, which of the following represents the law of readiness?
   A. Learning accompanied with satisfaction
   B. The more recent the practice, the more effective the performance
   C. Applying real-life experiences in realistic settings
   D. A person can learn when physically and mentally ready to learn

_____ 6. According to the laws of learning, which of the following represents the law of effect?
   A. Learning accompanied with satisfaction
   B. The more recent the practice, the more effective the performance
   C. Applying real-life experiences in realistic settings
   D. A person can learn when physically and mentally ready to learn

_____ 7. According to the laws of learning, which of the following represents the law of intensity?
   A. Learning accompanied with satisfaction
   B. The more recent the practice, the more effective the performance
   C. Applying real-life experiences in realistic settings
   D. A person can learn when physically and mentally ready to learn

_____ 8. Within the cognitive domain of learning, which of the following is represented by the use of information?
   A. Knowledge
   B. Comprehension
   C. Application
   D. Analysis

_____ 9. Within the cognitive domain of learning, which of the following is represented by the ability to remember information acquired in the past?
   A. Knowledge
   B. Comprehension
   C. Application
   D. Analysis

_____ 10. Within the cognitive domain of learning, which of the following is represented by being able to understand the meaning of information?
   A. Knowledge
   B. Comprehension
   C. Application
   D. Analysis

_____ 11. Which of the following psychomotor domain levels is the earliest that occurs and is defined as simply watching the skill or activity being performed?
   A. Observation
   B. Imitation
   C. Manipulation
   D. Precision

_____ 12. Which of the following psychomotor domain levels is the highest level of performance and is represented by the ability to perform multiple skills correctly?
   A. Manipulation
   B. Precision
   C. Articulation
   D. Naturalization

_____ 13. Which of the domains of learning is often referred to as the hardest to achieve or see results in?
   A. Cognitive
   B. Psychomotor
   C. Affective
   D. Forced learning

_____ 14. If the instructor asks a firefighter to complete a skill using a ground ladder by performing each of the steps required to raise the ladder, he or she is said to be training in which domain?
   A. Cognitive
   B. Psychomotor
   C. Affective
   D. Forced learning

_____ 15. Which theorist developed the Laws of Learning?
   A. Abraham Maslow
   B. Benjamin Bloom
   C. Edward Thorndike
   D. Alan Brunacini

Matching

Match the correct learning domain with its category where learning takes place.

- A. Affective
- B. Cognitive
- C. Psychomotor

_____ 1. Knowledge
_____ 2. Physical use of knowledge
_____ 3. Attitudes, emotions, or values

Match the level of cognitive learning with its proper definition.

- D. Knowledge
- E. Comprehension
- F. Application
- G. Analysis
- H. Synthesis
- I. Evaluation

_____ 4. Integration of information as a whole
_____ 5. Using the information
_____ 6. Remembering knowledge acquired in the past
_____ 7. Understanding the meaning of the information
_____ 8. Breaking the information into parts to help understand all the information
_____ 9. Using standards and criteria to judge the value of the information

Match the classification of learning disability with its short definition.

- J. Dysphasia
- K. Dyscalculia
- L. Dyspraxia
- M. Attention-deficit/hyperactivity disorder (ADHD)

_____ 10. A chronic level of inattention and impulsive actions
_____ 11. Difficulty with math subjects
_____ 12. Lack of physical coordination of motor skills
_____ 13. Inability to write, spell, or place words in the right order

True/False

If you believe the statement to be more true than false, write the letter *T* in the space provided. If you believe the statement to be more false than true, write the letter *F*.

_____ 1. Learning occurs in either formal or informal settings.
_____ 2. In the psychomotor domain, each level builds on the previous one.
_____ 3. A visual learner as described in the VAK (Visual, Auditory, and Kinesthetic characteristics) model is someone who prefers information to be displayed in words.
_____ 4. There are three levels of affective learning.
_____ 5. The use of teams is not a helpful method of instruction in classroom settings.
_____ 6. A student with ADHD may be protected by the Americans with Disabilities Act.
_____ 7. The acquisition of new values is an example of affective learning.
_____ 8. Knowledge may be defined as remembering past learning.
_____ 9. The term psychomotor refers to the use of the brain and the senses to tell the body what to do.
_____ 10. Using the appropriate teaching style for each learner in a class will not increase safety.

Fill-in
Read each item carefully, and then complete the statement by filling in the missing word(s).

1. _____ is a change in a person's ability to behave in certain ways.

2. Learning takes _____ and _____.

3. Each person has an individual _____, which is the way in which he or she learns the most effectively. This is a key principle for instructors.

4. Bloom's classification system of learning domains is widely known as _____.

5. The law of _____ states that a person can learn when he or she is physically and mentally prepared to respond to instruction.

Short Answer
Complete this section with short written answers using the space provided.

1. Describe the laws and principles of learning.

   _____
   _____
   _____
   _____
   _____

2. Identify the three types of learning domains.

   _____
   _____
   _____
   _____
   _____

3. Define learning styles, and discuss the effects of learning styles.

   _____
   _____
   _____
   _____
   _____

Instructor Applications

Discuss your response to these instructor applications to assist you in developing your knowledge of your responsibilities as a fire service instructor.

1. Review the laws and principles of learning, and identify an instructional method that incorporates each of the laws and principles.

   _____
   _____
   _____
   _____
   _____
   _____

2. Make a list of potential students you will have in upcoming classes, and based on their knowledge, experience, and other instructional factors, identify the ways each of these student classes learns and how you can make the best applications to them for the class.

   _____
   _____
   _____
   _____
   _____
   _____

3. Review how you would try to increase your students' understanding of basic material by adjusting the lesson plan to their individual needs.

   _____
   _____
   _____
   _____
   _____
   _____

Workbook Activities

The following activities have been designed to help you. Your instructor may require you to complete some or all of these activities as a regular part of your instructor training program. You are encouraged to complete any activity that your instructor does not assign as a way to enhance your learning in the classroom.

Chapter Review

The following exercises provide an opportunity to refresh your knowledge of this chapter.

Multiple Choice
Read each item carefully, and then select the best response.

_____ 1. How many elements are included in a properly functioning communication process?
   A. 2
   B. 3
   C. 4
   D. 5

_____ 2. Which part of the communication process represents the way the message is transmitted between sender and receiver?
   A. Message
   B. Medium
   C. Feedback
   D. Encoding

_____ 3. In the classroom, what does the instructor represent in the communication process?
   A. Sender
   B. Message
   C. Medium
   D. Receiver

_____ 4. An acknowledgment of a radio communication between an incident commander and an officer by repeating back a message represents what part of the communication process?
   A. Message
   B. Medium
   C. Feedback
   D. Affirmation

Communication Skills for the Instructor

Chapter 5

_____ 5. In the classroom, what does the student represent in the communication process?
   A. Sender
   B. Message
   C. Medium
   D. Receiver

_____ 6. What surrounds the communication process and has a direct influence on learning?
   A. The environment
   B. The message
   C. The medium
   D. The communication style

_____ 7. What should be the format for written communications for outside the organization?
   A. Memo form
   B. SOG form
   C. Letter form
   D. E-mail form

_____ 8. What can hand gestures used by an instructor express?
   A. Aggression
   B. Enthusiasm
   C. Nervousness
   D. All of the above

_____ 9. Your communication style should be
   A. Fixed
   B. Situational
   C. Consistent
   D. According to department standard

_____ 10. Passive listening involves the use of what?
   A. Your eyes
   B. Electronic equipment
   C. Hearing aids
   D. Anticipating the next words to be spoken

Matching

Match the element of the communications process with its simple definition.

A. Sender
B. Message
C. Medium
D. Receiver
E. Feedback

_____ 1. The student
_____ 2. An acknowledgment or action that completes the communication process
_____ 3. How the message is transmitted
_____ 4. The initiator of the process
_____ 5. What is being communicated

True/False

If you believe the statement to be more true than false, write the letter *T* in the space provided. If you believe the statement to be more false than true, write the letter *F*.

_____ 1. In order for the communications chain to function properly, all of the links need to be attached to one another.
_____ 2. One of the best ways to develop and improve your communication skills is to practice them.
_____ 3. After you have mastered the communication process, the learning environment becomes less of a factor in the learning process.
_____ 4. People who talk are always good communicators.
_____ 5. Speaking softly may be an effective tool in communications.
_____ 6. A student's reading level should not influence the instructor's delivery of information or the rate of learning.
_____ 7. Group exercises and role playing can be effective methods of communication and instruction, especially with experienced students.
_____ 8. The basic rule for writing is to include the three *W*'s and the *H*.
_____ 9. By simply raising or lowering your voice, you can emphasize important points.
_____ 10. The fire service has its own unique language.

Fill-in

Read each item carefully, and then complete the statement by filling in the missing word(s).

1. In many learning situations, the _____ breeds _____ in the learning process.

2. Communication may be _____ percent what you say and _____ percent how you say it.

3. The two types of listening are _____ and _____.

4. In order to be an effective instructor, you must become an effective _____.

5. Most presentations for fire service instructors involve _____, _____, or _____.

Short Answer

Complete this section with short written answers using the space provided.

1. Describe the elements of the communication process.

   _____
   _____
   _____
   _____
   _____
   _____
   _____
   _____

2. Describe the role of communication in the learning process.

   _____
   _____
   _____
   _____
   _____
   _____
   _____
   _____

3. Compare the different types of communication.

___

Instructor Applications

Discuss your response to these instructor applications to assist you in developing your knowledge of your responsibilities as a fire service instructor.

1. Make a list of common distracters that you have witnessed when being taught by your instructors. Describe how they impaired the learning process for you.

___

2. Describe the traits of a person who you felt was a great communicator in the classroom.

3. List the types of news stories or political speeches that have held your interest, and determine why they did so.

Workbook Activities

The following activities have been designed to help you. Your instructor may require you to complete some or all of these activities as a regular part of your instructor training program. You are encouraged to complete any activity that your instructor does not assign as a way to enhance your learning in the classroom.

Chapter Review

The following exercises provide an opportunity to refresh your knowledge of this chapter.

Multiple Choice

Read each item carefully, and then select the best response.

_____ 1. What is typically the starting point for lesson plan development that the content of the lesson plan is driven by?
   A. Learning objectives
   B. Preparation step
   C. Application step
   D. Evaluation instruments

_____ 2. One of the primary reasons to use a lesson plan is to
   A. Provide direction for class outcome
   B. Organize all material in a logical order
   C. Identify the evaluation process
   D. All of the above

_____ 3. In the ABCD method of objective development, what does the C stand for?
   A. Consistent
   B. Concise
   C. Confident
   D. Condition

_____ 4. How many components should be present in a properly prepared behavioral objective?
   A. 1
   B. 2
   C. 3
   D. 4

Chapter 6

Lesson Plans

_____ 5. What does the behavior of the learning objective describe?
   A. The student
   B. The situation in which the student will perform
   C. An observable and definable action
   D. How well the action is performed

_____ 6. What does the degree of the learning objective describe?
   A. The student
   B. The situation in which the student will perform
   C. An observable and definable action
   D. How well the action is performed

_____ 7. What does the audience of the learning objective describe?
   A. The student
   B. The situation in which the student will perform
   C. An observable and definable action
   D. How well the action is performed

_____ 8. What does the condition of the learning objective describe?
   A. The student
   B. The situation in which the student will perform
   C. An observable and definable action
   D. How well the action is performed

_____ 9. Which component of the lesson plan contains the main body or the outline of the material to be presented?
   A. Objective statement
   B. Lesson outline
   C. Resources and references
   D. Summary

_____ 10. In the four-step method of instruction, which step motivates and gets the student ready to learn?
   A. Preparation
   B. Presentation
   C. Application
   D. Evaluation

_____ 11. In the four-step method of instruction, which step gives the student the ability to demonstrate his or her comprehension of lesson material?
   A. Preparation
   B. Presentation
   C. Application
   D. Evaluation

_____ 12. Which step in the four-step method of instruction is considered to be the most important step?
   A. Preparation
   B. Presentation
   C. Application
   D. Evaluation

_____ 13. In terms of lesson plan development, which of the following represent the duties of the Instructor I in terms of adjusting a lesson plan from its original state?
   A. An Instructor I will alter the lesson plan.
   B. An Instructor I will modify the lesson plan.
   C. An Instructor I will create the lesson plan.
   D. An Instructor I cannot make any adjustments to a lesson plan.

_____ 14. What are the three lowest levels of cognitive objectives?
   A. Evaluation, synthesis, application
   B. Knowledge, comprehension, application
   C. Condition, behavior, standard
   D. Preparation, presentation, application

_____ 15. If a lesson plan requires the use of a Self-Contained Breathing Apparatus (SCBA) unit as a visual aid, where would this be information be located in the lesson plan?
   A. Learning objectives
   B. Materials needed
   C. References
   D. Application step

Matching

Match the part of the lesson plan with its simple definition.

A. Lesson title
B. Level of instruction
C. Behavioral objectives
D. Instructional materials
E. Lesson outline
F. References/resources
G. Lesson summary
H. Assignment

_____ 1. The review of the lesson plan
_____ 2. Homework or reading assignments
_____ 3. The description of what the lesson is about
_____ 4. Listing of props and instructional aids
_____ 5. The main body of the lesson plan
_____ 6. Identification of where lesson content was derived from
_____ 7. The specific outcomes of the lesson plan
_____ 8. Identification of the students who should be in the class

Chapter 6 Lesson Plans   33

Match the name of each step of the four-step method of instruction with its simple definition.

I. Preparation
J. Presentation
K. Application
L. Evaluation

_____ 9. The students will apply the knowledge presented in the class
_____ 10. The body and actual delivery of the lesson plan
_____ 11. The motivational step that gets the student ready to learn
_____ 12. The measure of students' knowledge or skills that have been delivered

True/False

If you believe the statement to be more true than false, write the letter *T* in the space provided. If you believe the statement to be more false than true, write the letter *F*.

_____ 1. It is not necessary to present all elements of a behavioral objective in order.
_____ 2. The Fire Service Instructor I cannot modify a lesson plan objective that identifies the number of fire fighters who will perform an evolution.
_____ 3. The Fire Service Instructor I cannot modify the lesson plan and course materials to meet the needs of the students and their learning styles.
_____ 4. A lesson plan should be reviewed to make sure it meets local Standard Operating Procedures (SOPs).
_____ 5. The Fire Service Instructor II is responsible for creating lesson plans.
_____ 6. A job performance requirement is converted into a learning objective when developing a lesson plan.
_____ 7. A knowledge-based objective requires the student to have the ability to solve problems or apply information.
_____ 8. Only a Fire Service Instructor II will present a prepared lesson plan.
_____ 9. The evaluation step of the four-step method of instruction ensures that students correctly acquired knowledge and skills typically through testing.
_____ 10. During the presentation step of the instructional process, the instructor prepares the student to learn through various motivational techniques.

Fill-in

Read each item carefully, and then complete the statement by filling in the missing word(s).

1. The Fire Service Instructor I should not _____ the content or the _____ of a lesson plan.

2. The Fire Service _____ will develop the content of a lesson plan.

3. A _____ is the goal that is achieved through attainment of a knowledge or skill.

4. You must identify the _____ in a lesson plan because the student must be able to understand the material.

5. The Instructor I may _____ a lesson plan to fit students' needs, whereas the Instructor II may _____ the lesson plan to change the fundamental content of it.

Short Answer

Complete this section with short written answers using the space provided.

1. List the components of learning objectives.

2. List and describe the parts of a lesson plan.

3. List the four-step method of instruction.

Instructor Applications

Discuss your response to these instructor applications to assist you in developing your knowledge of your responsibilities as a fire service instructor.

1. Using a sample lesson plan provided in class, identify each component.

   _____
   _____
   _____
   _____
   _____
   _____
   _____

2. Using the same lesson plan, identify what parts of the lesson plan would be adjusted to accommodate special needs of students.

   _____
   _____
   _____
   _____
   _____
   _____
   _____

3. Identify a locally developed lesson plan and determine whether all of the required elements of it are present.

   _____
   _____
   _____
   _____
   _____
   _____
   _____

Workbook Activities

The following activities have been designed to help you. Your instructor may require you to complete some or all of these activities as a regular part of your instructor training program. You are encouraged to complete any activity that your instructor does not assign as a way to enhance your learning in the classroom.

Chapter Review

The following exercises provide an opportunity to refresh your knowledge of this chapter.

Multiple Choice

Read each item carefully, and then select the best response.

_____ 1. The makeup of your audience based on race, gender, economics, or ethnic origin is referred to as what?
   A. Demographics
   B. Background information
   C. Diversity
   D. Democracy

_____ 2. In general, there are _____ types of fire departments?
   A. Two
   B. Three
   C. Five
   D. Seven

_____ 3. If students in a class that you are preparing to teach are not at the level of knowledge or understanding for the material you are to present, you should do what?
   A. Not teach the lesson plan as assigned and teach something else
   B. Call the lead instructor and ask them what to do
   C. Modify the lesson plan to meet the students' needs but teach the lesson plan as assigned
   D. Rewrite the lesson plan and change the lesson objectives

_____ 4. Which of the following is a *disadvantage* to a hollow square seating arrangement?
   A. Students can see each other.
   B. It is an ideal setup for small-group discussion.
   C. It allows students the opportunity to develop ideas.
   D. Students can easily be distracted or have off subject conversations.

Chapter 7

The Learning Environment

_____ 5. The best seating arrangement for conducting a demonstration of donning an SCBA would be what?
   A. Large U-shaped arrangement
   B. Hollow square
   C. Small V-shape
   D. "T" arrangement

_____ 6. One of the more often overlooked areas of a learning environment is what?
   A. Lighting
   B. Room temperature
   C. Noise distractions
   D. All of the above

_____ 7. Which of the following statements is *incorrect*?
   A. Eliminating noise distractions outside is very difficult.
   B. Informal training takes place in every environment.
   C. Small-group activities have limited use in the fire service.
   D. Classroom lighting is an important part of the learning process.

_____ 8. The greatest concern in the outdoor classroom is what?
   A. Noise
   B. Weather
   C. Safety
   D. Distractions

_____ 9. When asking students to introduce themselves at the beginning of a training session, the instructor should do what?
   A. Be aware of rude comments such as "I have to be here."
   B. Take notes as to why a student is attending the course.
   C. Listen for hints of expectations.
   D. All of the above are correct.

_____ 10. Prerequisites are used for all of the following *except*?
   A. Determine successful completion of a course.
   B. Ensure that students entering a course are at a certain level of knowledge.
   C. Satisfy NFPA requirements.
   D. Limit attendance to only students that pass an entrance exam.

_____ 11. The *highest* level of student success in the classroom environment as tied to Maslow's hierarchy is what?
   A. A warm learning environment
   B. The student successfully completing the course assignment
   C. The student having a feeling of safety in the learning environment
   D. The student being allowed to fail without worry of punishment

_____ 12. (1) Asking students to turn off cell phones before beginning instruction is not appropriate in today's world. (2) Limiting student access to the class until a break in the lecture is an acceptable method for controlling the classroom environment.
   A. Both statements are correct.
   B. Both statements are incorrect.
   C. Statement (1) is correct. Statement (2) is incorrect.
   D. Statement (1) is incorrect. Statement (2) is correct.

_____ 13. The basic or first level of Maslow's hierarchy of needs is what?
   A. A warm learning environment
   B. The student successfully completing the course assignment
   C. The student having a feeling of safety in the learning environment
   D. The student being allowed to fail without worry of punishment

_____ 14. What is the biggest difference between a class that is conducted indoors versus outdoors?
   A. Lighting
   B. Noise
   C. Weather
   D. Setup needs

_____ 15. Every fire department has its own unofficial way of doing business, and this is referred to as a department's what?
   A. Makeup
   B. Culture
   C. Policy
   D. History

Matching

Match each of the terms in the left column to the appropriate definition in the right column.

_____ 1. Learning environment
_____ 2. Demographics
_____ 3. Instructor I
_____ 4. Instructor II
_____ 5. Large U shape
_____ 6. Prerequisites
_____ 7. Classroom environment
_____ 8. NFPA 1403
_____ 9. Evaluation
_____ 10. Contingency plan

A. Standard on Live Fire Training
B. Responsible for setting up the classroom prior to giving a presentation
C. Factors that influence a person's behavior based on gender, age, or marital status
D. Used to demonstrate skills for a class
E. The perfect environment that satisfies all of Maslow's hierarchy of needs
F. Plan B
G. Environment when ideas are shared and learning is achieved
H. Develop evaluation tools for use in a course
I. Tool used to determine success of a student's increase in knowledge
J. Tool used to determine a student's starting point

True/False

If you believe the statement to be more true than false, write the letter *T* in the space provided. If you believe the statement to be more false than true, write the letter *F*.

_____ 1. Instructor level one is responsible for developing lesson plans.
_____ 2. Instructor level two determines instructor-to-student ratio.
_____ 3. Age is considered part of demographic information.
_____ 4. Offensive language in the classroom is normal and acceptable because of the nature of the fire service.
_____ 5. Typically, students are not concerned with course objectives.
_____ 6. It is acceptable to modify course objectives based on information obtained during the introduction of the class.
_____ 7. Content of the course material is more important than the classroom setup.
_____ 8. Breaks during classroom instruction should be limited to one every 2 hours.
_____ 9. There are no safety issues during classroom-only instruction.
_____ 10. Some of the greatest lessons learned from the fire ground take place on a tailboard in a truck bay after an incident.

Fill-in
Read each item carefully, and then complete the statement by filling in the missing word(s).

1. Department _____ is created from time, makeup of membership, and location of the agency.

2. A _____ training center with tables and chairs puts students in the mind frame that students _____ should match the location where they are.

3. A well-prepared instructor will always have a _____ plan in the event of bad weather.

4. Having the right number of _____, in the right position, doing the right things, with an emphasis on _____ makes the learning environment complete.

5. If you hear an off-color joke and _____, you are _____.

Short Answer
Complete this section with short written answers using the space provided.

1. Describe what encompasses the word "demographics."

   _____
   _____
   _____
   _____

2. Describe what should be considered in creating a learning environment.

   _____
   _____
   _____
   _____

3. Describe how Maslow's hierarchy of needs can be adapted into creating a learning environment.

   _____
   _____
   _____
   _____

Instructor Applications

Discuss your response to these instructor applications to assist you in developing your knowledge of your responsibilities as a fire service instructor.

1. Review an upcoming class you are scheduled to teach, and look at the demographics of the audience to see whether it changes how you prepare for this course.

2. Review your department's policy on instructor behavior or conduct. If your department does not have one, suggest the development of one and offer to help in its development.

3. Review your department's policy on classroom safety and training ground safety. Determine when it was last updated, and determine how encompassing it is.

Workbook Activities

The following activities have been designed to help you. Your instructor may require you to complete some or all of these activities as a regular part of your instructor training program. You are encouraged to complete any activity that your instructor does not assign as a way to enhance your learning in the classroom.

Chapter Review

The following exercises provide an opportunity to refresh your knowledge of this chapter.

Multiple Choice

Read each item carefully, and then select the best response.

_____ 1. What is a key element of most presentations?
   A. Multimedia tools
   B. Learning environment
   C. Clear speaker
   D. Room arrangement

_____ 2. Which of the following statements is *correct* concerning today's new fire fighters?
   A. New fire fighters do not know how to use today's technology.
   B. New fire fighters were raised using technology in the learning environment.
   C. New fire fighters like lectures and simulations.
   D. New fire fighters lack technology skills.

_____ 3. When using a PowerPoint presentation, you should do all of the following *except*?
   A. Do not read from the slides.
   B. Do not stand in front of the screen.
   C. Turn the lights off so everyone can read the slides.
   D. Have a backup plan in case the system does not work.

_____ 4. Slide projectors use _____-mm slides.
   A. 32
   B. 45
   C. 30
   D. 35

Chapter 8

Training Today: Multimedia Applications

_____ 5. What is the largest drawback to the use of slide projectors?
   A. Cost of the system
   B. Producing the slides
   C. Taking pictures for use in the presentation
   D. Using a slide projector

_____ 6. Which of the following is an *advantage* of a PowerPoint presentation?
   A. The ability to embed video and audio into your presentation
   B. Reading from the slide if you forget your lecture notes
   C. Overloading the students' senses with information so they will not ask questions
   D. Using technology to increase the learning curve

_____ 7. All of the following are advantages of videotapes except what?
   A. They are inexpensive.
   B. There is a wide selection available for use.
   C. They are copyrighted material.
   D. They can be viewed on any television with a VCR.

_____ 8. When purchasing a digital camera, you should buy one with a minimum of _____ megapixels.
   A. 10
   B. 21
   C. 8
   D. 4

_____ 9. Which of the following is an online learning platform?
   A. Angel
   B. MODEL
   C. Whiteboard
   D. WebbingGT

_____ 10. Which one of the following devices has been used for over 40 years?
   A. Computer
   B. Overhead projector
   C. Slide projector
   D. LCD projector

_____ 11. Which of the following is an example of a hand-held personal computer?
   A. ACB
   B. iPod
   C. PDA
   D. LCD

_____ 12. When creating PowerPoint slides, a rule is that headings should be _____ font, and the body text should be _____ font.
   A. 42, 32
   B. 56, 38
   C. 28, 36
   D. 38, 28

_____ 13. An important part of modern maintenance is having what?
   A. Antivirus protection
   B. Extra light bulbs
   C. Cleaning solutions
   D. Router cleaner

_____ 14. All of the following are reasons for using multimedia for presentations *except*?
   A. Multimedia presentations allow students to see, hear, and in some cases touch the learning process.
   B. Multimedia entertains students and takes pressure off the instructor.
   C. Multimedia simplifies complex theories into an understandable format.
   D. Multimedia keeps today's students interested in learning.

_____ 15. The biggest issue facing instructors who want to use multimedia is what?
   A. The cost
   B. The students' learning level
   C. What media is appropriate for the material to be taught
   D. The instructors

Matching

Match each of the terms in the left column to the appropriate definition in the right column.

_____ 1. PowerPoint presentations
_____ 2. Wireless microphone, transmitter, and receiver
_____ 3. Keystoning
_____ 4. Project a page from a textbook or a map or diagram
_____ 5. Adobe Flash Player
_____ 6. Computer-based training
_____ 7. Classroom and Internet learning combined
_____ 8. Virtual academy
_____ 9. Satellite programming
_____ 10. Trench training in a controlled environment

A. Hybrid training
B. Visual projector
C. File extension "swf"
D. Fire and Emergency Training Network
E. Audio sound system
F. Blackboard
G. PPT
H. National Emergency Training Center
I. Improper slide projector setup
J. Models

True/False

If you believe the statement to be more true than false, write the letter *T* in the space provided. If you believe the statement to be more false than true, write the letter *F*.

_____ 1. Handouts are considered a form of media.

_____ 2. Chalkboards are considered more "healthy" than erasable boards.

_____ 3. An advantage of easel pads is the ability to save the information and post it around a classroom.

_____ 4. Clip art files are where you find pictures form newspapers that can be copied and passed out to a class.

_____ 5. Quicktime media programs are used to develop PowerPoint slides.

_____ 6. PowerPoint slides should support a lecture or presentation.

_____ 7. Ghosting is a term that refers to faint shadows that appear to the right of letters or numbers on a slide.

_____ 8. Norton and McAfee are examples of software that is used to embed video into a PPT.

_____ 9. The proper use of multimedia methodology to the learning process will increase the student's ability to understand the information.

_____ 10. When using multimedia during a presentation, an instructor should have contingency plans A and B and C.

Fill-in

Read each item carefully, and then complete the statement by filling in the missing word(s).

1. Multimedia tools come in a _____ of types and are intended for multiple uses.

2. A disadvantage of multimedia is the potential for _____ on technology as part of your instruction.

3. An advantage of a _____ is the ability to take printed material and load it onto your computer.

4. A _____ allows students to download lesson material and listen to it on an iPod.

5. A _____ can be used to teach incident command in a controlled classroom environment.

Short Answer

Complete this section with short written answers using the space provided.

1. What can you do as an instructor to learn to use media for your presentations?

   _____
   _____
   _____
   _____
   _____
   _____
   _____
   _____

2. What are three methods for troubleshooting projected media?

   _____
   _____
   _____
   _____
   _____
   _____
   _____
   _____

3. What are two advantages of using multimedia during a presentation?

   _____
   _____
   _____
   _____
   _____
   _____
   _____
   _____

Instructor Applications

Discuss your response to these instructor applications to assist you in developing your knowledge of your responsibilities as a fire service instructor.

1. Experiment with a PowerPoint presentation. Which color schemes work best for dark classrooms? Which work best for bright classrooms? Learn how to change a PowerPoint presentation quickly so that you can instantly adapt to existing conditions.

   _____
   _____
   _____
   _____
   _____
   _____
   _____

2. Modify a PowerPoint presentation by adding text, photos, or other illustrations to clarify the message.

   _____
   _____
   _____
   _____
   _____
   _____
   _____

3. Select a learning platform such as an online or distance learning course. Discover what the benefits of that platform are for your students.

   _____
   _____
   _____
   _____
   _____
   _____
   _____

Workbook Activities

The following activities have been designed to help you. Your instructor may require you to complete some or all of these activities as a regular part of your instructor training program. You are encouraged to complete any activity that your instructor does not assign as a way to enhance your learning in the classroom.

Chapter Review

The following exercises provide an opportunity to refresh your knowledge of this chapter.

Multiple Choice

Read each item carefully, and then select the best response.

_____ 1. In a training situation, you as the instructor should
   A. Always put safety ahead of course objectives.
   B. Always put course objectives ahead of outcome assessments.
   C. Always keep cost under budget to ensure continued funding.
   D. Always put instructor safety ahead of trainee safety.

_____ 2. One of the 16 Fire Fighter Initiatives is to reduce line of duty deaths by:
   A. 25 percent over the next 10 years
   B. 50 percent over the next 10 years
   C. 70 percent over the next 5 years
   D. 50 percent over the next 5 years

_____ 3. When it comes to safety, fire service instructors should
   A. Delegate safety to others
   B. Follow NFPA 1404 during Live Fire Training
   C. Lead by example in everything they do
   D. Expect others to do as they say but not what they do

_____ 4. Safety during training begins
   A. With the attitude of the fire chief
   B. In development of training objectives
   C. Once you hit the training ground
   D. With the classroom environment

Chapter 9

Safety During the Learning Process

_____ 5. An important rule is that the level of Personal Protective Equipment required for an emergency is
   A. The level of PPE that should be worn during training
   B. A minimum standard set by the NFPA
   C. Not required during training
   D. Determined by each incident

_____ 6. Which NFPA standard establishes the policy for rehabilitation during training?
   A. NFPA 1403
   B. NFPA 1584
   C. NFPA 1500
   D. NFPA 1710

_____ 7. Failure by the instructor to correct improper skill performance on the training ground can lead to what?
   A. The fire fighter passing the course material
   B. The company officer sending a letter to the training chief
   C. Unsafe performance during a real life incident
   D. Failing the next evaluation

_____ 8. Which of the following would *not* be considered high-risk training?
   A. Swift water training
   B. Live fire evolutions
   C. Liquefied Petroleum Gas fire evolutions
   D. Trench simulator training

_____ 9. Which of the following Incident Command System positions is mandated to be staffed during a live fire evolution?
   A. Safety officer
   B. Water supply officer
   C. Rehab officer
   D. Operations officer

_____ 10. According to NFPA 1403, the person responsible for planning and coordinating all training activities is the
   A. Incident commander
   B. Instructor in charge
   C. Safety officer
   D. Lead instructor

_____ 11. When selecting instructors to assist in the training process, you should consider all of the following *except*
   A. Their attitude toward safety
   B. Their skills and qualifications
   C. Their gender or race
   D. Their presentation skills

_____ 12. (1) Tragedies during training are usually single events that occur without warning. (2) Accidents or injuries that occur during training are usually a series of small problems that build into an event or accident.
   A. Statement 1 is correct. State 2 is incorrect.
   B. Statement 1 is incorrect. Statement 2 is correct.
   C. Both statements are correct.
   D. Both statements are incorrect.

_____ 13. During an accident investigation, it is important to
   A. Control information released by the department
   B. Be selective in who is interviewed
   C. Seek answers not blame
   D. Determine who is at fault

_____ 14. (1) Students have a responsibility for their own safety. (2) Students have a responsibility for meeting all of the prerequisites for a course.
   A. Statement 1 is correct. State 2 is incorrect.
   B. Statement 1 is incorrect. Statement 2 is correct.
   C. Both statements are correct.
   D. Both statements are incorrect.

_____ 15. Who is responsible for staying current on laws, regulations, and standards that affect fire fighter training?
   A. Fire chief
   B. Safety officer
   C. City attorney
   D. Training officer

Matching

Match each of the terms in the left column to the appropriate definition in the right column.

_____ 1. Reducing fire fighter line of duty deaths
_____ 2. Initiative number five
_____ 3. Classroom safety
_____ 4. Setting up ladders and climbing to a roof
_____ 5. Use of incident command, rehabilitation, and use of full PPE during training
_____ 6. Supervise other instructors and the use of safety policies
_____ 7. Providing a learning environment that protects the emotional side of the fire fighter
_____ 8. Student-to-instructor ratio during high-risk training
_____ 9. Following lesson outline and adhering to department safety policies
_____ 10. Practicing knots in a classroom environment

A. Fire Service Instructor I responsibilities
B. Five to one
C. Hands-on training
D. Department training SOPs
E. Fire Service Instructor II responsibilities
F. Sixteen Life-Safety Initiatives
G. Identify exits and procedures in the event of an emergency
H. Low-risk training
I. Train to national standards applicable to all fire fighters based on duties assigned
J. Hidden hazards during the training process

True/False

If you believe the statement to be more true than false, write the letter *T* in the space provided. If you believe the statement to be more false than true, write the letter *F*.

_____ 1. On average, 10 fire fighters die each year during training activities.
_____ 2. The Sixteen Fire Fighter Life-Safety Initiatives are mandatory requirements.
_____ 3. It is acceptable to modify PPE requirements during training.
_____ 4. One of the main causes of injuries during training is the failure to follow SOPs.
_____ 5. The strongest message you send as an instructor is what you say in the classroom.
_____ 6. Safety in the training environment should be stated in the lesson plan.
_____ 7. The use of an unsafe chain saw is better than not using one at all.
_____ 8. It is better to stress or push a new recruit during training than to let him or her fail on the fire ground.
_____ 9. All prerequisite training must be met prior to participating in a live fire exercise.
_____ 10. During live fire training, the instructor in charge fills the role of the IC.

Fill-in

Read each item carefully, and then complete the statement by filling in the missing word(s).

1. All fire fighters must be _____ to stop unsafe practices.

2. Use available _____ wherever it can produce higher levels of health and _____.

3. The fire service has been _____ in developing methods and practices to accomplish drills and training.

4. An adequate number of _____ must be present to conduct training safely.

5. Fire-ground _____ translate to the _____.

Short Answer

Complete this section with short written answers using the space provided.

1. How can you adapt the 16 Fire Life-Safety Initiatives into your departments training activities?

2. How can you teach safety to a profession that by its nature operates in a dangerous environment?

3. How do you develop a safety culture?

Instructor Applications

Discuss your response to these instructor applications to assist you in developing your knowledge of your responsibilities as a fire service instructor.

1. Review local policies relating to safety in training.

   _____
   _____
   _____
   _____
   _____
   _____
   _____
   _____

2. Read NFPA 1403, *Standard on Live Fire Training*, and review the sample checklists used for conducting live fire training exercises.

   _____
   _____
   _____
   _____
   _____
   _____
   _____
   _____

3. Research fire fighter line-of-duty deaths in training accidents, and review the recommendations of the investigators regarding ways to prevent similar occurrences.

   _____
   _____
   _____
   _____
   _____
   _____
   _____
   _____

Workbook Activities

The following activities have been designed to help you. Your instructor may require you to complete some or all of these activities as a regular part of your instructor training program. You are encouraged to complete any activity that your instructor does not assign as a way to enhance your learning in the classroom.

Chapter Review

The following exercises provide an opportunity to refresh your knowledge of this chapter.

Multiple Choice

Read each item carefully, and then select the best response.

_____ 1. All of the following are reasons for testing *except*?
   A. To justify the purchase of new equipment
   B. To measure student attainment of learning objectives
   C. To enhance and improve training programs
   D. To identify and determine weaknesses and gaps in training programs

_____ 2. Which of the following testing tools is not used in the fire service very often?
   A. Written
   B. Oral
   C. Performance
   D. Psychomotor

_____ 3. A test that measures the students' ability to perform a task is an example of a(n)
   A. Oral test
   B. Written test
   C. Skills evaluation
   D. Cognitive test

_____ 4. The term used to describe how well a test item measures what the developer intended it to measure is
   A. Reliability
   B. Currency
   C. Dependability
   D. Validity

Chapter 10

Evaluating the Learning Process

_____ 5. The lowest form of test validity is
   A. Face
   B. Technical content
   C. Job content
   D. Currency

_____ 6. _____ means the test item measure the knowledge in a consistent manner.
   A. Reliability
   B. Currency
   C. Dependability
   D. Validity

_____ 7. Which written test is the most widely used in the fire service?
   A. True/false
   B. Multiple choice
   C. Matching
   D. Fill in the blank

_____ 8. The most important advantage of the multiple choice test item is that it can be used
   A. To weight a test.
   B. To distract the unlearned student from the correct answer.
   C. To measure higher mental functions such as reasoning and judgment.
   D. To ask valid and invalid questions.

_____ 9. When developing a matching test tool, you should do all of the following *except*?
   A. Include only one correct match for each item
   B. Arrange statements and responses in random order
   C. Keep all items on the same page of the test booklet
   D. Have the same number of questions and answers

_____ 10. Which of the following testing tools is *not* considered to be objective?
  A. True/false
  B. Multiple choice
  C. Matching
  D. Essay

_____ 11. Which testing method is the single most important method for determining the competency for actual task performance?
  A. Performance
  B. Written
  C. Oral
  D. Affective

_____ 12. Who must give permission for the release of a students' test scores?
  A. The instructor
  B. The student
  C. The chief
  D. The training officer

_____ 13. What is the most correct method for releasing a student's test score?
  A. Post scores with social security numbers on the bulletin board.
  B. Read off the names and scores in the classroom during class time.
  C. One-on-one basis in a private office.
  D. You cannot release student test scores but must send them to the department for release.

_____ 14. When proctoring a written exam, you should do all of the following *except*?
  A. Arrive early and set up the classroom.
  B. Space students away from each other as much as possible.
  C. Monitor students at all times, never leaving the testing room.
  D. Give out answers to questions that you feel are incorrect.

_____ 15. If you observe a student cheating during a written exam, you should
  A. Document the activity but let the student continue taking the test.
  B. Ask the student to leave and tear up the test.
  C. Follow department policy that addresses this issue.
  D. Challenge the student during the test and ask him or her to leave.

Matching

Match each of the terms in the left column to the appropriate definition in the right column.

_____ 1. Written testing tool
_____ 2. "Given a hand tool, explain in detail how to clean the tool"
_____ 3. Don an SCBA within 1 minute
_____ 4. Test item developed by subject matter expert
_____ 5. Develop and analyze evaluation instruments to ensure that learning objectives are tested
_____ 6. Proctor a written exam to a class
_____ 7. Test item measures what it is intended to measure
_____ 8. Not the correct answers on a multiple-choice exam
_____ 9. Testing tool with a 50/50 chance of getting the answer correct
_____ 10. Assigning relative point values to key responses

A. Instructor I responsibility
B. Validity
C. Technical content validity
D. Oral test
E. Essay question matrix
F. Short-answer essay
G. True/false question
H. Distracters
I. Performance evaluation
J. Instructor II responsibility

True/False

If you believe the statement to be more true than false, write the letter *T* in the space provided. If you believe the statement to be more false than true, write the letter *F*.

_____ 1. Essay questions are the easiest to evaluate and grade.

_____ 2. Test questions developed by the instructor are not valid.

_____ 3. Test banks developed by a company are more valid than test banks developed by an instructor.

_____ 4. Proctoring a written exam is the same as evaluating a skills exam.

_____ 5. Testing tools should be developed following APA guidelines.

_____ 6. As an instructor, you should never stop a skills exam or evaluation.

_____ 7. Providing feedback to students is just as important as teaching a lesson.

_____ 8. A written feedback form could be considered a legal document of training performance.

_____ 9. Qualitative analysis is used to determine the acceptability of a question.

_____ 10. Currency of information being asked in a question is as important as is the face value of a question.

Fill-in

Read each item carefully, and then complete the statement by filling in the missing word(s).

1. A sound _____ program allows you to know whether a student is progressing in the learning process.

2. A _____ can be made up of eight types of test items.

3. The purpose of _____ is to determine whether test items are functioning as desired and to eliminate, correct, or modify those test items.

4. The _____ test item is efficient for measuring the application of procedures for starting an apparatus.

5. A _____ will address what an instructor should do in the event of a student cheating on an exam.

Short Answer

Complete this section with short written answers using the space provided.

1. Describe some of the common problems with testing.

2. Describe the four forms of test-item validity.

3. Describe the purpose for analyzing test results.

Instructor Applications

Discuss your response to these instructor applications to assist you in developing your knowledge of your responsibilities as a fire service instructor.

1. Review a recent written examination and compare the student responses to the answer key. Identify any test questions that were either too easy (every student answered the question correctly) or that had poor success (there is a high percentage of failures).

   _____
   _____
   _____
   _____
   _____
   _____

2. Review your local authority policy, and practice posting test scores and student completion records.

   _____
   _____
   _____
   _____
   _____
   _____

3. Develop a strategy to improve student performance based on a poor evaluation result.

   _____
   _____
   _____
   _____
   _____
   _____

Workbook Activities

The following activities have been designed to help you. Your instructor may require you to complete some or all of these activities as a regular part of your instructor training program. You are encouraged to complete any activity that your instructor does not assign as a way to enhance your learning in the classroom.

Chapter Review

The following exercises provide an opportunity to refresh your knowledge of this chapter.

Multiple Choice

Read each item carefully, and then select the best response.

_____ 1. What was created during the Wingspread Conference in 1971?
   A. Professional qualifications for fire fighters
   B. Standards for evaluating fire fighters
   C. Fire prevention activities for the fire service
   D. The three E's of fire prevention

_____ 2. When is a fire service instructor evaluated?
   A. During class by another instructor
   B. By the students during a break in the break area
   C. At the end of class during a formal course evaluation
   D. All of the above are correct

_____ 3. (1) Instructor qualifications are set by the authority having jurisdiction. (2) Instructor qualifications are set by the NFPA 1041 standard.
   A. Statement 1 is correct. State 2 is incorrect.
   B. Statement 1 is incorrect. Statement 2 is correct.
   C. Both statements are correct.
   D. Both statements are incorrect.

_____ 4. Which of the following would *not* be evaluated while doing an evaluation of a new instructor in your training division?
   A. Clothing or attire of the instructor
   B. Occasional references to his or her own department when appropriate
   C. Use of projected media and transitions
   D. Following the lesson plan as outlined

Chapter 11

Evaluating the Fire Service Instructor

_____ 5. _____ evaluations are intended to refine the instructor's delivery and better prepare him or her to deliver a course.
   A. Summative
   B. Formative
   C. Conclusive
   D. Promotional

_____ 6. Which type of evaluation tool is usually used at the end of the class to determine curriculum strengths and weaknesses?
   A. Summative
   B. Formative
   C. Conclusive
   D. Promotional

_____ 7. Which of the following statements is *incorrect* regarding student course evaluations?
   A. Students might use course evaluations to take a shot at the instructor.
   B. Students make comments about course length that are opinions versus constructive criticism.
   C. Course evaluations should be the key source of information for merit increases.
   D. Course evaluation tools should be used to improve the course material.

_____ 8. One of the primary functions of a course evaluation tool is to
   A. Determine students' attitude about the course
   B. Use the information from the evaluations to discipline the instructor
   C. Justify increasing the budget for the training division
   D. Identify the instructor's weaknesses and then develop a plan to help him or her improve

_____ 9. Which level of certification requires Fire Service Instructor I as a prerequisite?
   A. Fire Fighter I
   B. Fire Officer I
   C. Fire Investigator
   D. Public Fire and Life Safety-Educator

_____ 10. Which type of evaluation can be the most critical on an instructor?
   A. Student evaluation
   B. Peer evaluation
   C. Self-evaluation
   D. Promotional evaluation

_____ 11. When evaluating another instructor, the first thing you should do on arriving at the classroom is to
   A. Meet with the instructor and let them know why you are there and what you will be doing.
   B. Slip into the back of the classroom, and make no contact with the instructor you will be evaluating.
   C. Visit with students in the class, and ask them how they like the class prior to starting your evaluation.
   D. Arrive ahead of class time and set up a video camera to videotape the instructor without his or her knowledge.

_____ 12. During the evaluation of an instructor during training ground activities, the highest priority should be
   A. Ensuring that activities match the lesson plan
   B. Ensuring that safety is always observed
   C. Verifying that the instructor-to-student ratio is within department policy
   D. Ensuring the organization of the training evolutions

_____ 13. After evaluating an instructor, the next step should be
   A. Leave as quickly as possible, and compile your notes.
   B. Talk with the instructor in front of the class with students still present.
   C. Meet with the instructor one on one, and discuss your evaluation.
   D. Ask the instructor for his or her thoughts; add them to your notes, and then leave.

_____ 14. When discussing your evaluation of an instructor, you should do all of the following *except*?
   A. Be objective in your comments
   B. Be honest and friendly
   C. Make some suggestions to improve observed weaknesses
   D. Be blunt and to the point

_____ 15. All of the following could be areas of measurement found on an evaluation tool *except* what?
   A. Did the class start on time?
   B. Did you like your classmates?
   C. Was the instructor properly attired?
   D. Was the room temperature comfortable?

Matching

Match each of the terms in the left column to the appropriate definition in the right column.

_____ 1. Fire Service Instructor Professional Qualifications standard
_____ 2. Average fire fighter deaths during training activities per year (2000–2005)
_____ 3. Improving the fire service instructor's performance
_____ 4. Evaluation completed at the end of a course for information on the course
_____ 5. Evaluation scale found on a student survey
_____ 6. Create an evaluation tool to evaluate another instructor
_____ 7. Is evaluated by students, supervising instructors
_____ 8. Fire Officer Professional Qualifications standard
_____ 9. Attire, eye contact, voice strength, and subject knowledge
_____ 10. Standard on Fire Department Occupational Safety and Health

A. NFPA 1500
B. NFPA 1021
C. Fire Service Instructor II
D. Formative evaluation
E. 10
F. NFPA 1041
G. Evaluation tool
H. Strongly agree or disagree
I. Summative evaluation
J. Fire Service Instructor I

True/False

If you believe the statement to be more true than false, write the letter T in the space provided. If you believe the statement to be more false than true, write the letter F.

_____ 1. Fire service evaluations are always tied to promotions or merit increases.
_____ 2. On occasion, it is acceptable to deviate from the lesson plan to reinforce a point.
_____ 3. Performing instructor evaluations is for Fire Service Instructor II only.
_____ 4. Fire department policy can exceed the NFPA requirements.
_____ 5. Instructor evaluations are limited to classroom presentation.
_____ 6. Preparation for an evaluation begins with reviewing the lesson plan.
_____ 7. Evaluating an instructor is as important as evaluating the student.
_____ 8. Word of mouth evaluations should be considered an objective source of information.
_____ 9. NFPA 1500 is the basis for all professional qualifications.
_____ 10. Feedback to an instructor should be given with the same confidentiality as a student.

Fill-in

Read each item carefully, and then complete the statement by filling in the missing word(s).

1. The lesson plan is the fire service instructor's _____.

2. Evaluating the fire service instructor on a(n) _____ or more frequent basis ensures that the instructor does not become _____.

3. During the evaluation, look at the fire service instructor's entire _____.

4. Closely scrutinizing a fire service instructor's adherence to student _____ guidelines should be paramount.

5. Remind the fire service instructor that it is his or her obligation to remain a _____ at all times.

Short Answer

Complete this section with short written answers using the space provided.

1. Describe your department's policy or procedure in respect to fire service instructor evaluations.

2. Describe the evaluation process.

3. Describe the role for providing feedback to the fire service instructor.

Instructor Applications

Discuss your response to these instructor applications to assist you in developing your knowledge of your responsibilities as a fire service instructor.

1. Review the content of a course evaluation form, and identify those elements that relate to fire service instructor qualities as well as those elements that relate to the features of the course.

   _____
   _____
   _____
   _____
   _____
   _____
   _____

2. Develop a list of desirable fire service instructor qualities. Rank these qualities in order based on which create the best learning environment.

   _____
   _____
   _____
   _____
   _____
   _____
   _____

3. Consult available standard personnel evaluations, and identify any features that could be used to evaluate a fire service instructor during a training session.

   _____
   _____
   _____
   _____
   _____
   _____
   _____

Workbook Activities

The following activities have been designed to help you. Your instructor may require you to complete some or all of these activities as a regular part of your instructor training program. You are encouraged to complete any activity that your instructor does not assign as a way to enhance your learning in the classroom.

Chapter Review

The following exercises provide an opportunity to refresh your knowledge of this chapter.

Multiple Choice

Read each item carefully, and then select the best response.

_____ 1. A critical part of the fire department's efforts to protect its members is the
   A. Training officer and instructors
   B. The fire chief
   C. The company officer
   D. The mayor and city council

_____ 2. As a supervisor of a fire department training program, you are responsible for all of the following *except*
   A. Following department policies in regards to safety
   B. Dealing with city council members who want to cut your training budget
   C. Use of personnel accountability during live fire evolutions
   D. Assignment of qualified instructors to teach classes

_____ 3. Which of the following is *not* an area of consideration when scheduling training?
   A. EMS recertification hours
   B. Hazardous materials OSHA training
   C. CERT refresher training
   D. Department-mandated HR policies

_____ 4. The majority of training that takes place at a career fire department is
   A. On the job
   B. Rookie school
   C. Fire academy training
   D. In-service drills

Chapter 12

Managing the Training Team

_____ 5. Which of the following is *not* an in-service training category?
   A. Skill/knowledge competence
   B. Skill/knowledge development
   C. Skill/knowledge maintenance
   D. Skill/knowledge improvement

_____ 6. (1) When developing a training schedule, you have to consider all areas of the department such as inspections and technical rescue.
   (2) Under OHSA, technical rescue teams are responsible for their own skills maintenance and do not follow department policy.
   A. Statement 1 is correct. State 2 is incorrect.
   B. Statement 1 is incorrect. Statement 2 is correct.
   C. Both statements are correct.
   D. Both statements are incorrect.

_____ 7. When developing a training schedule, you should consider all of the following questions *except*?
   A. Who must attend the training session?
   B. Will this training impact mutual aid agreements?
   C. Who will instruct the training session?
   D. What resources are needed for the training session?

_____ 8. Training requirements could come from all of the following agencies *except*?
   A. OSHA
   B. Insurance Service Office (ISO)
   C. Underwriters Lab
   D. State EMS bureau

_____ 9. The first step in creating a master training calendar is to
   A. Identify ongoing activities that will affect the training schedule.
   B. Identify specialized training needs.
   C. Complete an agency needs assessment of mandatory training.
   D. Compare training requirements proposed by your agency's human resource department.

_____ 10. The third step in the budgetary process is what?
   A. Identification of needs and required resources
   B. Preparation of a budget request
   C. Adoption of an approved budget
   D. Government and public review process

_____ 11. When purchasing supplies for a course, the Fire Service Instructor II should do all of the following *except* what?
   A. Know how to keep budget requests under the maximum bid amount.
   B. Know his or her department's bid process.
   C. Identify vendors that have the material you want to purchase.
   D. Know what your budget is and how to issue a purchase order.

_____ 12. All of the following are factors seen in a poor training program *except*?
   A. Inconsistency of information between fire service instructors
   B. Following lesson plans and department policies
   C. Failure to adhere or enforce safety practices
   D. Unclear instructions or learning outcomes

_____ 13. One criterion that outweighs all others in the evaluation of the fire service instructor is what?
   A. Presentation skills
   B. Adherence to department policy
   C. Safety
   D. Proper instructor-to-student ratio

_____ 14. When posting a training notice, you should include all of the following information *except* what?
   A. What the training subject is
   B. What level of PPE or classroom attire is appropriate
   C. Where the training will take place
   D. What the budget was for the course

_____ 15. According to NFPA 1500, all training should be based on what?
   A. Job performance requirements (JPRs)
   B. OSHA
   C. ISO
   D. State mandates

Matching

Match each of the terms in the left column to the appropriate definition in the right column.

_____ 1. New approach to fire-fighting operations is adopted by a department

_____ 2. Training intended to develop performance baselines for core duties and functions

_____ 3. Training intended to correct poor performance or errors observed on the fire ground

_____ 4. Supplemental training schedule

_____ 5. Agency training needs assessment

_____ 6. An itemized summary of estimated or intended revenues and expenditures

_____ 7. Income from taxes, bonds, grants, or fees

_____ 8. Money spent for goods or services

_____ 9. July 1 through June 30

_____ 10. Products that are manufactured by only one vendor

A. Revenues
B. Single-source product
C. Skill/knowledge improvement
D. Expenditures
E. Skill/knowledge maintenance
F. Budget
G. Fiscal year
H. Contingency plan B
I. Skill/knowledge development
J. Basis on which department training is established

True/False

If you believe the statement to be more true than false, write the letter *T* in the space provided. If you believe the statement to be more false than true, write the letter *F*.

_____ 1. Quality assurance is an important aspect of managing a training program.

_____ 2. It is acceptable practice to allow instructors to present lesson plans using their own style.

_____ 3. In most departments, it is best not to publish a training schedule that helps to ensure better attendance.

_____ 4. Training requirements for volunteer fire fighters is not the same as for career fire fighters.

_____ 5. Mandatory training is required only in OSHA states.

_____ 6. NFPA enforces OSHA.

_____ 7. The master training schedule should be the responsibility of one person.

_____ 8. The Fire Service Instructor I is responsible for the development of the training budget.

_____ 9. Successful departments have a training schedule that is consistent and easy to understand.

_____ 10. The master training schedule is the basis for a department's training budget.

Fill-in

Read each item carefully, and then complete the statement by filling in the missing word(s).

1. Training and education are tools used by fire departments to improve their efficiency in operations and in fire fighter _____.

2. Surveys of training programs often reveal that the delivery of training suffers because of poor _____ of the training _____.

3. Advanced policies may also detail the responsibilities of those who _____, supervised, and _____ in the training.

4. OSHA has specific training requirements that should be considered, covering topics ranging from _____ to _____.

5. Successful _____ requires justification of the amount of money being requested.

Short Answer

Complete this section with short written answers using the space provided.

1. Describe how to schedule an instructional session.

   _____
   _____
   _____
   _____
   _____
   _____
   _____
   _____

2. Describe the budget process for your agency.

   _____
   _____
   _____
   _____
   _____
   _____
   _____
   _____

3. Describe how to schedule instructors for a training session.

   _____
   _____
   _____
   _____
   _____
   _____
   _____
   _____

Instructor Applications

Discuss your response to these instructor applications to assist you in developing your knowledge of your responsibilities' as a fire service instructor.

1. Review your department's training policy, and identify management-related responsibilities that must be completed by the Fire Service Instructor II.

   _____
   _____
   _____
   _____
   _____
   _____
   _____
   _____

2. Obtain a copy of the training budget to review expenditures and possible revenue sources.

   _____
   _____
   _____
   _____
   _____
   _____
   _____
   _____

3. Review your department's training schedules. Outline the process for determining training subjects, fire service instructor assignments, and the resources needed as part of the overall training program.

   _____
   _____
   _____
   _____
   _____
   _____
   _____
   _____

Workbook Activities

The following activities have been designed to help you. Your instructor may require you to complete some or all of these activities as a regular part of your instructor training program. You are encouraged to complete any activity that your instructor does not assign as a way to enhance your learning in the classroom.

Chapter Review

The following exercises provide an opportunity to refresh your knowledge of this chapter.

Multiple Choice

Read each item carefully, and then select the best response.

_____ 1. Which of the following staff positions is in a prime place to influence the next generation of fire fighters?
   A. Chief
   B. Company officer
   C. Training officer/instructor
   D. None of the above

_____ 2. Which of the following is one method for the fire service instructor to establish his or her creditability?
   A. Teaching a lot of classes
   B. Publishing his or her work in trade journals or books
   C. Attending instructor conferences
   D. Earning a college degree

_____ 3. What sets the fire service leader apart from the average fire fighter?
   A. Having a desire to improve and be better
   B. Always looking for the easy way out of situations
   C. Never volunteering for extra duty or assignments
   D. Following traditions

_____ 4. One of the first skills you should learn to do to become a leader in the fire service is to
   A. Attend a lot of conferences.
   B. Read as many articles on the Internet or from magazines.
   C. Develop an extensive network of friends.
   D. Learn time management skills and set goals.

Chapter 13

The Learning Process Never Stops

_____ 5. What is a potential problem or drawback to information found on the Internet?
  A. Cost of Internet access
  B. The accuracy of the information found on the Internet
  C. The time it takes to download information
  D. Copyright issues with material found on the Internet

_____ 6. What is the value of reading articles that contain information from other professions or disciplines outside the fire service?
  A. It might help you find a new profession other than the fire service.
  B. Nothing at all—you should focus on your profession.
  C. It helps to make you a more "well-rounded" person and gives you a more of a diverse perspective.
  D. It breaks up the routine of reading just fire service material.

_____ 7. What is the educational requirement currently in place to gain entrance into the Executive Fire Officer Program offered by the National Fire Academy?
  A. Bachelors degree
  B. Associate degree
  C. Masters degree
  D. There is no educational requirement in place.

_____ 8. One of the most important benefits for attending major conferences is to
  A. Network and develop friendships
  B. Learn new skills
  C. Take ideas from others and use them as your own
  D. Be away from your work for a period of time

_____ 9. How many 2-week courses make up the Executive Fire Officer Program?
   A. 3
   B. 5
   C. 4
   D. 2

_____ 10. The process of identifying others to train, mentor, and coach is known as
   A. Succession planning
   B. Replacement training
   C. Progressive retirement
   D. Retooling

_____ 11. Which one of the following is *not* a concept in training your replacement?
   A. Mentoring
   B. Sharing
   C. Coaching
   D. Formalizing

_____ 12. The first step in selecting a person to develop as your replacement is
   A. Mentoring
   B. Identifying
   C. Coaching
   D. Asking

_____ 13. When mentoring a new protégé, you should do all of the following *except* what?
   A. Be a sounding board and provide feedback.
   B. Be patient and allow time for development.
   C. Mold him or her into your image as you see it.
   D. Share experiences with your protégé, both good and bad.

_____ 14. The process of letting individuals develop solutions on their own is known as what?
   A. Mentoring
   B. Coaching
   C. Teaching
   D. Challenging

_____ 15. What is something all fire service instructors leave with their departments after they have retired?
   A. Their lesson plans
   B. Their training records
   C. Their attitude
   D. Their legacy of professionalism

Matching

Match each of the terms in the left column to the appropriate definition in the right column.

_____ 1. Giving to others the "tricks to the trade"
_____ 2. Allowing the protégé to teach a class and then giving him or her feedback
_____ 3. Acting as a sounding board to new ideas and suggestions
_____ 4. National model for degree programs
_____ 5. Four-year program taught at the NFA/NETC in Maryland
_____ 6. National organization that is committed to the development of the instructor
_____ 7. Process of developing contacts with a common interest
_____ 8. EFOP requirement starting FY 2009 for entrance into program
_____ 9. Process that brings together education, training, and experience
_____ 10. Example of professionalizing the fire service

A. Networking
B. ISFSI
C. Mentoring
D. Bachelors degree
E. FESHE
F. Sharing
G. Publishing an article in a magazine
H. Coaching
I. EFOP
J. National Professional Development Model

True/False

If you believe the statement to be more true than false, write the letter *T* in the space provided. If you believe the statement to be more false than true, write the letter *F*.

_____ 1. Lifelong learning is a process, not a stopping point.
_____ 2. Information is only as good as its source.
_____ 3. All information found on the Internet is accurate and correct.
_____ 4. Time management is a lifelong skill for professional development.
_____ 5. College education is the only way the fire service can develop into a profession.
_____ 6. Networking is used only for political gain and empowerment.
_____ 7. It is a Fire Service Instructor I and Fire Service Instructor II requirement to develop instructor networks.
_____ 8. The EFOP and National Professional Development model is the same thing.
_____ 9. Everyone selected for mentoring works out and succeeds.
_____ 10. Coaching requires positive feedback to be successful.

Fill-in

Read each item carefully, and then complete the statement by filling in the missing word(s).

1. The benefit from reading material outside the fire service profession is that it helps you develop into a more _____ individual and professional.

2. More departments have begun requiring an advanced degree for _____, such as a _____ degree for battalion chief.

3. _____ is a powerful tool in your toolbox that can open the door to sharing or collaborating.

4. Keeping your _____ sharp and current is just as important as reading magazines or attending a conference.

5. The true mark of a great leader is a willingness to plan to be _____.

Short Answer

Complete this section with short written answers using the space provided.

1. Describe the ideas involved in lifelong learning.

2. Describe the value and importance of time management.

3. Create a list of two or three individuals who could be developed for succession planning for your position.

Instructor Applications

Discuss your response to these instructor applications to assist you in developing your knowledge of your responsibilities' as a fire service instructor.

1. Identify at least three statewide organizations and three national organizations that are available to assist all levels of instructors in their professional development.

   _____
   _____
   _____
   _____

2. Review your local or state certification process to learn which affiliations or correlations exist in accordance with the Fire and Emergency Services Higher Education (FESHE) program, discussed in this chapter, "The 16 Firefighter Life-Safety Initiatives" developed by the National Firefighters Foundation, and NFPA standards relating to training and education.

   _____
   _____
   _____
   _____

3. Review the JPRs that are specific to your next level of instructor certification, and identify opportunities to experience learning in those areas.

   _____
   _____
   _____
   _____

4. Select an instructor or instructor candidate who has specific talents or background that can be used to improve a particular course. Urge that person to become involved in the training program.

   _____
   _____
   _____
   _____

Fire Service Instructor II

Student Application Package

Developing Objectives, Lesson Plans, and Course Materials

Fire Service Instructor II
Student Applications

Assignment # _____   Due Date: _____   Pass: _____   Fail: _____

Student Application Package Overview and Guidelines

<u>Purpose:</u> To provide the Instructor II candidate with opportunity to develop a practice instructional package for presentation to course instructor(s) and fellow students completing the job performance requirements of the fire service instructor II.

<u>Skill:</u> Manage instructional resources and develop instruction materials used to conduct a specific lesson topic to a student audience. (NFPA 1041; 5.2, 5.3, 5.4, 5.5)

<u>Competencies:</u> Prepare and present a complete lesson plan and supporting instructional material to a student audience completing the following tasks:

- Identify a class subject area for illustrative or manipulative training
- Develop three cognitive objectives for subject area
- Develop three psychomotor objectives for subject area
- Prepare a properly formatted lesson plan based on subject area choice
- Develop appropriate media for cognitive presentation
- Acquire instructional resources for psychomotor presentation
- Develop an assignment sheet or information sheet on lesson plan information
- Use a test planning sheet to plan evaluation methods for each objective
- Develop a manipulative skills evaluation checklist
- Develop a written examination using multiple types of evaluation questions
- Develop a class/instructor evaluation form
- Deliver all prepared materials to student audience (15 to 20 minutes)

Fire Service Instructor II
Student Applications

Assignment # _____   Due Date: _____   Pass: _____   Fail: _____

Task: Construct a properly formatted cognitive performance objective. (NFPA 1041; 5.3.2)

Instructions: Write three properly formatted objectives in the spaces provided below.

BEHAVIOR: A statement that describes the task activity, knowledge, or attitude being sought.

Examples: List classes of fire, describe vertical ventilation, or identify ladder parts.

1. _____
2. _____
3. _____

CONDITION: A statement of the circumstances under which the outcome will be observed or measured. Should relate as closely as possible to the time, limits, materials, or equipment that the fire fighter will be confronted with when performing the task.

Examples: Given a written exam, from a selection of pictures, or given a sample lesson plan.

1. _____
2. _____
3. _____

DEGREE: The level or quality of the outcome that when achieved will identify acceptable attainment of the task. The standard should reflect acceptable job performance.

Examples: Within 30 seconds, with 100% accuracy, or according to department SOP.

1. _____
2. _____
3. _____

Write a complete performance objective by assembling the three components listed previously here. Make sure that the objective is specific, observable, and measurable.

1. _____
   _____
2. _____
   _____
3. _____
   _____

Fire Service Instructor II
Student Applications

Assignment # _____   Due Date: _____   Pass: _____   Fail: _____

Task: Construct a properly formatted psychomotor performance objective. (NFPA 1041; 5.3.2)

Instructions: Write three properly formatted objectives in the spaces provided below.

BEHAVIOR: A statement that describes the task activity, knowledge, or attitude being sought.

Examples: Don SCBA, raise the extension ladder, or demonstrate a straight hose roll.

1. _____
2. _____
3. _____

CONDITION: A statement of the circumstances under which the outcome will be observed or measured. Statement should relate as closely as possible to the time, limits, materials, or equipment that the fire fighter will be confronted with when performing the task.

Examples: While wearing full protective clothing, given a ladder, or given a 1½-, 2½-, or 3-inch hose.

1. _____
2. _____
3. _____

DEGREE: The level or quality of the outcome that when achieved will identify acceptable attainment of the task. The standard should reflect acceptable job performance.

Examples: Within 30 seconds, with 100% accuracy, or according to department SOP.

1. _____
2. _____
3. _____

Write a complete performance objective by combining the above three components. Make sure the objective is specific, observable, and measurable.

1. _____
   _____
2. _____
   _____
3. _____
   _____

Fire Service Instructor II
Student Applications

Assignment # _____ Due Date: _____ Pass: _____ Fail: _____

Task: Develop a properly formatted lesson plan for presentation. (NFPA 1041; 5.3.2)

Instructions: Assemble the components for a lesson plan to present to students in class.

LESSON TITLE: Title hints at the topic and gives learner some idea of what to expect.

_____

TYPE OF PRESENTATION: Identify if presentation is cognitive or psychomotor level.

☐ Cognitive Domain   ☐ Psychomotor Domain

LEARNING OBJECTIVES: The objective describes what learners will accomplish.

1. _____
   _____

2. _____
   _____

3. _____
   _____

TIME FRAME: Gives the instructor a time frame for completing the lesson.

_____

LEVEL OF INSTRUCTION: Describes the level of desired outcome from the lesson. Are the fire fighters expected to know the material to a basic knowledge level or to the analysis level?

_____

MATERIALS NEEDED: List all of the materials, including quantity needed to teach the lesson.

Examples: Number of handouts, equipment needed, videotapes, and so forth.

_____
_____
_____

Fire Service Instructor II

Student Applications

Assignment # _____ Due Date: _____ Pass: _____ Fail: _____

Task: Develop a properly formatted lesson plan for presentation. (Continued)

REFERENCES: List the references and resources that were used to develop the lesson. Include page numbers where appropriate.

_____

_____

PREPARATION (step 1): This section reminds the instructor to provide the learners with some type of reason why they need to know this information. How or where will they use this new skill? Motivate the student to pay attention and learn from your presentation.

Examples: Statistics, case study, recent incidents, video segment.

_____

_____

_____

PRESENTATION (step 2): Outline the lesson. This segment should be in the same order in which the information will be presented. Write in template in a logical order, step by step, and so forth.

Note: This is a sample format; your lesson plan may be longer/shorter than space provided.

I.
- A.
- B.
- C.
    1.
    2.

II.
- A.
- B.
    1.
    2.
    3.

III.
- A.
- B.
- C.
- D.

IV.
- A.
- B.
- C.
    1.
    2.
    3.

Fire Service Instructor II

Student Applications

Assignment # _____   Due Date: _____   Pass: _____   Fail: _____

Task: Develop a properly formatted lesson plan for presentation. (Continued)

APPLICATION (step 3): Whenever new information is given or a new skill is taught, it must be flowed with an opportunity for the fire fighter to apply the new knowledge. Detail what will be used to allow fire fighters to apply what was learned.

   Examples: Ask direct questions, provide an exercise, practice the skill, and role play.

_____
_____
_____
_____
_____

EVALUATION (step 4): Evaluate the student's performance. This section should tie into the learning objective. Using the behavior, condition, and standard components of the objective determines how you will measure successful achievement of the objectives. In this segment, you do not need to include your test; only indicate how you will measure the fire fighters.

   Examples: Written exam, practical skills test, oral exam, or written exercise.

_____
_____

LESSON SUMMARY: Review the main points of the lesson. This helps to clarify any confusion before dismissal of the class. In most cases, you can review the major topic headings in your class outline.

_____
_____
_____
_____
_____

ASSIGNMENTS: What does the fire fighter need to do to prepare for the next lesson?

   Examples: Read Chapter x, complete the prefire plan, or practice donning the SCBA five times at your station.

_____

Fire Service Instructor II
Student Applications

Assignment # _____ Due Date: _____ Pass: _____ Fail: _____

Task: Develop instructional media to support lesson plan and select audiovisual aids. (NFPA 1041; 5.3.2)

Instructions: Develop instructional media to support the delivery of your lesson plan and identify the audiovisual aids that will be used to deliver your media.

SELECT INSTRUCTIONAL MEDIA: The instructional media should maximize the transfer of knowledge and skills within the time allotted. The instructional media should meet the following criteria:

- ☐ Appeals to the senses
- ☐ Saves time
- ☐ Relevant to the course objectives
- ☐ Appropriate for the size and interaction of the class
- ☐ Appropriate to the pace of learning
- ☐ Practiced prior to class

EXAMPLES:

- ☐ PowerPoint slides
- ☐ DVD or CD-ROM
- ☐ Diagrams, charts, graphs
- ☐ Models, props

ASSIGNMENT: From your lesson plan, develop your presentation media to support your delivery of the lesson outline to the class participants.

GENERAL GUIDELINES: If using PowerPoint, the following suggestions may help you complete your presentation:

- Limit the words on your slides to support key points; don't write in sentences.
- Select background, animations, and graphics that do not distract from your presentation.
- Be consistent in font, color, size, and style of your text.
- PRACTICE, PRACTICE, PRACTICE.
- Save your presentation in the correct format for delivery, and make sure that all versions of the program are compatible and that you have a backup plan.
  - Use a flash drive backed up with a CD-ROM.

Fire Service Instructor II
Student Applications

Assignment # _____ Due Date: _____ Pass: _____ Fail: _____

Task: Develop a student learning resource for use during practice presentation. (NFPA 1041; 5.3.2)

Instructions: Develop a student learning resource for use during your practice presentation to support the understanding of course objectives.

Suggested content guide for information sheet.

CREATE A TITLE OR LIST THE JOB: Indicates the subject area and relates the title to the lesson.

LIST A BEHAVIORAL OBJECTIVE: The objective describes what the information sheet is designed to accomplish.

EXPLAIN IMPORTANCE OF THE INFORMATION: Briefly describe the information, and explain how the information will help the student. Motivate the student to read the information.

PRESENT THE INFORMATION: Make the information easy to read. Use charts, graphs, tables, and so forth to help explain the information. Use a separate sheet if necessary.

SUMMARIZE THE INFORMATION OR PROVIDE ADDITIONAL RESOURCES/ASSIGNMENTS: Prepare the student for evaluation step, and make follow-up or additional informational resources available to the students.

EXAMPLES:
- ☐ Copy of a case summary
- ☐ Department SOP
- ☐ PowerPoint slides handout
- ☐ Fill in the blank or write-in handout
- ☐ Key point summary of terms/concepts

Fire Service Instructor II
Student Applications

Assignment # _____   Due Date: _____   Pass: _____ Fail: _____

Task: Develop a skill evaluation sheet for a psychomotor objective. (NFPA 1041; 5.5.2)

Instructions: Develop a student skill evaluation sheet for a psychomotor objective that covers all aspects of the performance skill being taught.

EVALUATION INFORMATION: Your form should include student name, ID number, date, and location, as applicable to your department recordkeeping requirements.

LIST THE SKILL (Title): Describes what the student will be tested on.

  Example: soft-sleeve hydrant hook-up

_____

BEHAVIORAL OBJECTIVE: Select a psychomotor objective, and write into form. This provides clarity to the student and instructor on what will be performed, how it will be done, and to what degree the task will be performed.

_____

_____

STUDENT INSTRUCTIONS: Briefly explain to the fire fighter the task that he or she is to perform. Give details of limitations, conditions, and time frames required of the student.

_____

_____

_____

INSTRUCTOR INSTRUCTIONS: Explain to the instructor how the test is to be conducted. Detail any limitations, conditions, time frames, and so forth that the instructor should be aware of.

_____

_____

_____

Fire Service Instructor II
Student Applications

Assignment # _____ Due Date: _____ Pass: _____ Fail: _____

Task: Develop a skill evaluation sheet for a psychomotor objective. (Continued)

DEVELOP LIST OF THE STEPS TO COMPLETE THE TASK/SKILL: This list consists of the step-by-step procedure of completing the task. Start with the first step in the process and end with the final objective outcome. (You may need to consult with manufacturer instructions to complete this step.)

- Indicate in the task area in the table below. Expand or reduce the table to fit your skill.

DEVELOP RATING SYSTEM: Determine how the fire fighter will be evaluated. If using a scale or point system, provide a description of what is passing.

- As a psychomotor skill, the most accepted practice is to have a pass-all evaluation to be successful. The student needs to complete all steps in order to be successful. Other measures may help separate performances of students.
  - Excellent
  - Good
  - Average
  - Fair
  - Poor/Fail

TASK	PASS	FAIL
1.		
2.		
3.		
4.		
5.		

COMMENTS: Provide a place for the instructor to comment on the fire fighter's performance. The instructor could make suggestions for improvement if the fire fighter failed the test.

_____

_____

_____

SIGNATURE OF EVALUATOR: _____

SIGNATURE OF STUDENT: _____

Fire Service Instructor II
Student Applications

Assignment # _____   Due Date: _____   Pass: _____   Fail: _____

Task: Develop a written evaluation instrument for course objectives. (NFPA 1041; 5.5.2)

Instructions: Develop a written evaluation for lesson objectives using multiple forms of evaluation measures.

NOTE: The test planning sheet will be used to help plan the characteristics of your test.

1. WRITE FOUR MULTIPLE CHOICE QUESTIONS: The questions should relate to the class objectives along with the lesson plan. **PROVIDE DIRECTIONS ON HOW TO COMPLETE THIS SEGMENT:**

  Example: Choose the correct answer.

  Directions: _____

  1. _____
     _____
     A. _____
     B. _____
     C. _____
     D. _____

  2. _____
     _____
     A. _____
     B. _____
     C. _____
     D. _____

  3. _____
     _____
     A. _____
     B. _____
     C. _____
     D. _____

  4. _____
     _____
     A. _____
     B. _____
     C. _____
     D. _____

Fire Service Instructor II
Student Applications

Assignment # _____ Due Date: _____ Pass: _____ Fail: _____

Task: Develop a written evaluation instrument for course objectives. (Continued)

2. **WRITE TWO TRUE/FALSE QUESTIONS:** The questions should relate to the class objectives along with the lesson plan. **PROVIDE DIRECTIONS ON HOW TO COMPLETE THIS SEGMENT:**

   Example: Choose the correct answer.

   Directions: _____

   5. _____
      _____

      A. True

      B. False

   6. _____
      _____

      A. True

      B. False

3. **WRITE TWO COMPLETION QUESTIONS:** The questions should relate to the class objectives along with the lesson plan. **PROVIDE DIRECTIONS ON HOW TO COMPLETE THIS SEGMENT:**

   Example: Complete the statement.

   Directions: _____

   7. xxxxxxxxx xxxxxxxx xx xxxx xxxxxxxx xx _____ xxxxx xxxx xxxxxxxxxx xxxxxx.

   8. xxxxxxxxx xxxxxxxx xx xxxx xxxxxxxx xx _____ xxxxx xxxx xxxxxxxxxx xxxxxx.

4. **WRITE ONE SHORT ANSWER QUESTION:** The questions should relate to the class objectives along with the lesson plan. **PROVIDE DIRECTIONS ON HOW TO COMPLETE THIS SEGMENT:**

   Example: Provide a brief explanation.

   Directions: _____

   9. _____
      _____
      _____
      _____

Fire Service Instructor II
Student Applications

Assignment # _____   Due Date: _____   Pass: _____   Fail: _____

Task: Develop a written evaluation instrument for course objectives. (Continued)

5. WRITE ONE MATCHING QUESTION WITH FOUR RESPONSES: The question should relate to the class objectives along with the lesson plan. **PROVIDE DIRECTIONS ON HOW TO COMPLETE THIS SEGMENT:**

Example: Match the term or item in column A with the appropriate response in column B.

Directions: _____

A	B
10. _____	A. _____
11. _____	B. _____
12. _____	C. _____
13. _____	D. _____
	E. _____

Fire Service Instructor II
Student Applications

Assignment # _____   Due Date: _____   Pass: _____   Fail: _____

Task: Construct a course and instructor evaluation form. (NFPA 1041; 5.5.3)

Instructions: Develop a course and instructor evaluation form for use by the student audience to review presentation delivery characteristics.

Survey Example: Include instructions for student completion.

    Mark your reaction by circling one of the following choices:
    SA—if you strongly agree
    A—if you moderately agree
    D—if you disagree
    SD—if you strongly disagree
    N—if you neither agree nor disagree

    SA A D SD N   I would take another course like this.
    SA A D SD N   I did not learn anything from this class.

Questionnaire Example: Include instructions for student completion.

    Was this course what you expected it to be? If no, why not?

    What areas could be shortened? Entirely eliminated?

Rating Sheet Example: Include instructions for student completion.

    Rating Scale: 5 = outstanding
                           4 = more than satisfactory
                           3 = satisfactory
                           2 = less than satisfactory
                           1 = poor

    Printed materials were well organized.   5 / 4 / 3 / 2 / 1
    Course was reasonable length.   5 / 4 / 3 / 2 / 1
    Classroom contained minimum distractions.   5 / 4 / 3 / 2 / 1

Fire Service Instructor II
Student Applications

Assignment # _____   Due Date: _____   Pass: _____   Fail: _____

Instructions: Write your objectives or objective number in the left column, and complete additional column information to ensure that your evaluation instrument is comprehensive.

Test Planning Sheet

Objective *Indicate Standard Reference* *(NFPA or Local)*	Objective Type *Cognitive or Psychomotor*	Evaluation Method <u>Written</u> <u>Practical</u> <u>Other</u> M.C. Individual Group T/F Team Exercise S/A Mult. Obj. Problem Fill-In Essay Matching	Course/Text/Class Reference	Test Question Number or Practical Evaluation Title

Fire Service Instructor II
Student Applications

Assignment # _____   Due Date: _____   Pass: _____   Fail: _____

Student Information Sheet for Instructor II

1. Class attendance requirements:

2. Required textbook is *Fire Service Instructor*, 1st edition, Jones and Bartlett Publishers.

3. *Fundamentals of Firefighting Skills*, 2nd edition, can be used to obtain lesson plan material.

4. Final class projects should be developed using format suggested in the *Student Applications Guidebook*.

5. The student shall develop three cognitive and three psychomotor objectives. Each objective shall be in the format of the behavior/conditions/degree method as presented in the class.

6. The student shall develop a lesson plan on a subject the student feels comfortable with. Each lesson plan must be based upon the objectives written for each domain. The lesson plan shall be in the format presented in class and shall contain all key items as presented in class. Each lesson plan will be limited to 15 to 20 minutes. Students will make three copies of their lesson plan.

7. The student shall develop an assignment sheet or an information sheet relating to an objective in his or her cognitive lesson plan.

8. The student shall develop an instructor/class presentation evaluation sheet. Copies will be made and distributed to the class while presentations are made. Students will evaluate each other.

9. The student shall develop a skill sheet based upon the psychomotor objective used in the lesson plan.

10. Based on the lesson plan objective(s), the student shall develop an examination. The examination shall contain the following types of questions: *At direction of course instructor, copies of sample examination may be required to be administered to fellow students.*

    - Four multiple choice
    - Two true and false
    - Two completion
    - One short answer
    - One matching (**must** contain a minimum of four items + one extra answer)

Fire Service Instructor II
Student Applications

Assignment # _____   Due Date: _____   Pass: _____   Fail: _____

Student Information Sheet for Instructor II (Continued)

11. The student shall furnish copies of the following for evaluators:

    - Evaluation sheet (student shall hand out additional forms to peers as above)
    - Assignment/information sheet for illustrated lecture
    - Skill sheet for psychomotor lesson
    - Examination
    - Lesson plan

12. The student must use at least one instructional aid during the course of the presentation.

Answer Key

Chapter 1

Multiple Choice

1. A   pg. 4
2. D   pg. 4
3. C   pg. 5
4. C   pg. 5
5. D   pg. 6
6. A   pg. 8
7. C   pg. 9
8. D   pg. 11
9. B   pg. 12
10. A   pg. 13
11. C   pg. 13
12. A   pg. 16

Matching

1. B   pg. 6
2. B   pg. 6
3. A   pg. 6
4. A   pg. 6
5. C   pg. 6
6. B   pg. 6
7. A   pg. 6
8. B   pg. 6
9. C   pg. 9
10. C   pg. 10
11. A   pg. 9
12. E   pg. 11
13. D   pg. 10

True/False

1. False   pg. 5
2. True   pg. 6
3. True   pg. 6
4. True   pg. 1
5. True   pg. 1
6. True   pg. 13
7. False   pg. 15
8. False   pg. 16
9. True   pg. 171
10. False   pg. 161

Fill-in

1. professional qualification standards   pg. 5
2. success, failure   pg. 13
3. Standards   pg. 15
4. Codes of ethics   pg. 16
5. accurately recorded   pg. 16

Short Answer

1. Manage the basic resources and the records and report essential to the instructional process.
   - Assemble course materials.
   - Prepare training records and report forms.

   Review and adapt prepared instructional materials.
   Deliver instructional sessions using prepared course materials.
   - Organize the classroom, laboratory, or outdoor learning environments.
   - Use instructional media and materials to present prepared lesson plans.
   - Adjust presentations to students' different learning styles, abilities, and behaviors.
   - Operate and use audiovisual equipment and demonstration devices.

   Administer and grade student evaluation instruments.
   - Deliver oral, written, or performance tests.
   - Grade students' oral, written, or performance tests.
   - Report test results.
   - Provide examination feedback to students (pg. 6).

2. Manage instructional resources, staff, facilities, and records and reports.
   - Schedule instructional sessions.
   - Formulate budget needs.
   - Acquire training resources.
   - Coordinate training recordkeeping.
   - Evaluate instructors.

   Develop instruction materials for specific topics.
   - Create lesson plans.
   - Modify existing lesson plans.

   Conduct classes using a lesson plan.
   - Use multiple teaching methods and techniques to present a lesson plan that the instructor has prepared.
   - Supervise other instructors and students during training.

   Develop student evaluation instruments to support instruction and evaluation of test results.
   - Develop student evaluation instruments.
   - Develop a class evaluation instrument.
   - Analyze student evaluation instruments (pg. 6).

3. Physical (i.e., lighting, temperature, and setup)
   Emotional (i.e., attitudes, comments, and learning abilities) (pg. 11)
4. Leader, mentor, coach, evaluator, teacher (pg. 8)
5. Leading by example, accountability, recordkeeping, knowledge power base, trust, and confidentiality (pg. 16–17)

Instructor Applications

Individual answers will depend on department operations and individual beliefs and impressions.

Chapter 2

Multiple Choice

1. D   pg. 25
2. C   pg. 26
3. A   pg. 29
4. B   pg. 31
5. A   pg. 32
6. D   pg. 31
7. A   pg. 32
8. A   pg. 26
9. C   pg. 31
10. D   pg. 34

Matching

1. B   pg. 24
2. A   pg. 24
3. C   pg. 24
4. F   pg. 24
5. D   pg. 24
6. E   pg. 24
7. I   pg. 26
8. K   pg. 26
9. H   pg. 26
10. J   pg. 26
11. G   pg. 26
12. O   pg. 34
13. N   pg. 34
14. M   pg. 34
15. L   pg. 34

True/False

1. True   pg. 29
2. False   pg. 29
3. True   pg. 33
4. False   pg. 24
5. True   pg. 34
6. True   pg. 34
7. False   pg. 33
8. True   pg. 33
9. True   pg. 33
10. True   pg. 33

Fill-in

1. harassing slurs   pg. 29
2. reasonable accommodation pg. 29
3. sexual advances   pg. 29
4. misfeasance   pg. 32
5. confidentiality   pg. 34

Short Answer

1. In cases where an injury occurs to a participant or observer of a training session, it is crucial to document what occurred, who was present, and what each person observed (pg. 34).

2. Personnel files usually include information such as date of birth, social security number, dependent information, and medical information.
    Hiring files include test scores, pre-employment physical reports, psychological reports, and personal opinions about the candidate.
    Disciplinary files include any report or document about an individual's disciplinary history and related reports (pg. 34).
3. This should include federal law, which is made up of statutory laws; state law, which is established by each state's legislature; and policies and procedures crafted by each individual fire department (pg. 24).
4. This includes physical or mental impairment that substantially limits one or more major life activities. Major life activities include functions such as caring for oneself, performing manual tasks, walking, seeing, hearing, speaking, breathing, learning, and working (pg. 26).
5. Works include written words, photographs, and some art work (pg. 33).

Instructor Applications

Individual answers will depend on department operations and individual beliefs and impressions.

Chapter 3

Multiple Choice

1. A pg. 42
2. C pg. 42
3. A pg. 42
4. B pg. 43
5. D pg. 43
6. B pg. 43
7. C pg. 46
8. A pg. 47
9. B pg. 49
10. C pg. 52
11. B pg. 52
12. B pg. 51

Matching

1. C pg. 43, 44
2. B pg. 43, 44
3. D pg. 43, 44
4. A pg. 43, 44
5. F pg. 44
6. G pg. 44
7. E pg. 44
8. K pg. 51
9. H pg. 51
10. M pg. 51
11. I pg. 51
12. L pg. 51
13. J pg. 51

True/False

1. True pg. 42
2. False pg. 51
3. True pg. 52
4. True pg. 50
5. True pg. 47
6. False pg. 44
7. True pg. 43
8. True pg. 49
9. False pg. 51
10. True pg. 51
11. True pg. 44
12. False pg. 44

Fill-in

1. motivation   pg. 42
2. generation x   pg. 47
3. short pause   pg. 50
4. 10 percent, 90 percent   pg. 51
5. formal education   pg. 46

Short Answer

1. Why the adult learner is in class, knowing that attendance is mandatory, why the students must know the information, how the student will learn the material (pg. 43)
2. Distraction free, centered on student participation and comfort, easy to see and hear (pg. 43, 44)
3. Effective learning is a natural process, three types of learners, motivation as a key factor, many similarities and differences in learners
4. To deal with the *monopolizer*, make certain to call on other students during classroom discussions.

    With the *historian*, carefully guide the discussion back to the main topic. Another way to redirect real historians is by pairing them up with struggling students. Artificial historians are almost always distracting. Assertively redirecting the discussion back to the lesson plan generally corrects this issue. If not, take the student aside and discuss the need for the class to stay focused on the lesson plan.

    For the *day dreamer*, try to draw such students into the class by using direct questioning techniques and making maximum eye contact.

    For the *expert*, you must corral the expert and may even have to discuss his or her contributions to the class during a private break (pg. 51, 52).
5. Baby boomers, involve senior members with experience and knowledge. Generation X, many questions and seek instant gratification. Generation Y, spoiled with little mechanical but great comprehensive abilities (pg. 47–50)

Instructor Applications
Individual answers will depend on department operations and individual beliefs and impressions.

Chapter 4
Multiple Choice
1. D   pg. 58
2. C   pg. 62
3. A   pg. 63
4. B   pg. 59
5. D   pg. 59
6. A   pg. 58
7. C   pg. 59
8. C   pg. 62
9. A   pg. 62
10. B   pg. 62
11. A   pg. 63
12. D   pg. 63
13. C   pg. 65
14. B   pg. 63
15. C   pg. 58

Matching
1. B   pg. 62
2. C   pg. 62
3. A   pg. 62
4. H   pg. 62
5. F   pg. 62
6. D   pg. 62
7. E   pg. 62
8. G   pg. 62
9. I   pg. 62
10. M   pg. 68
11. K   pg. 68
12. L   pg. 68
13. J   pg. 68

True/False
1. True   pg. 58
2. True   pg. 63
3. False pg. 67
4. False pg. 63
5. False pg. 60
6. True   pg. 67
7. True   pg. 63
8. True   pg. 62
9. True   pg. 62
10. False pg. 65

Fill-in
1. Learning   pg. 58
2. time, patience   pg. 58
3. learning style   pg. 65
4. Taxonomy of Learning   pg. 62
5. readiness   pg. 58

Short Answer
1. The law of readiness: A person can learn why physically and mentally he or she is ready to respond to instruction.
    The law of exercise: Learning is an active process that exercises both the mind and the body.
    The law of effect: Learning is most effective when it is accompanied by or results in a feeling of satisfaction, pleasantness, or reward (internal or external) for the student.
    The law of association: In the learning process, the learner compares the new knowledge with his or her existing knowledge base.
    The law of recency: Practice makes perfect, and the more recent the practice, the more effective the performance of the new skill or behavior.
    The law of intensity: Real-life experiences are more likely to produce permanent behavioral changes, making this type of learning very effective (pg. 58, 59).
2. Cognitive—knowledge, psychomotor—physical use of knowledge, affective—attitudes, emotions, or values (pg. 62)

3. Each student has a different *learning style*—that is, a way in which he or she learns most effectively (pg. 65).

Instructor Applications

Individual answers will depend on department operations and individual beliefs and impressions.

Chapter 5

Multiple Choice

1. D   pg. 76
2. B   pg. 76
3. A   pg. 76
4. C   pg. 77
5. D   pg. 77
6. A   pg. 76
7. C   pg. 83
8. D   pg. 78
9. B   pg. 86
10. A pg. 78

Matching

1. D   pg. 77
2. E   pg. 77
3. C   pg. 77
4. A   pg. 77
5. B   pg. 77

True/False

1. True   pg. 76
2. True   pg. 78
3. True   pg. 76
4. False pg. 80
5. True   pg. 80
6. False pg. 80
7. True   pg. 86
8. False pg. 83
9. True   pg. 80
10. True pg. 81

Fill-in

1. enthusiasm, enthusiasm   pg. 80
2. 10, 90   pg. 77
3. active, passive   pg. 78
4. communicator   pg. 80
5. informative speeches, lectures   pg. 84

Short Answer

1. The basic *communication process* consists of five elements: the sender, the message, the medium, the receiver, and feedback.
2. It is the most important element of the learning process. Takes many formal and informal forms, instructors guide flow and pace.
3. Verbal, nonverbal, and written

Instructor Applications

Individual answers will depend on department operations and individual beliefs and impressions.

Chapter 6

Multiple Choice

1. A pg. 93
2. D pg. 92
3. D pg. 93
4. D pg. 93
5. C pg. 93
6. D pg. 93
7. A pg. 93
8. B pg. 93
9. B pg. 94
10. A pg. 98
11. C pg. 98
12. C pg. 99
13. B pg. 103
14. B pg. 104
15. B pg. 99

Matching

1. G pg. 94
2. H pg. 94
3. A pg. 94
4. D pg. 94
5. E pg. 94
6. F pg. 94
7. C pg. 94
8. B pg. 94
9. K pg. 98, 99
10. J pg. 98, 99
11. I pg. 98, 99
12. L pg. 98, 99

True/False

1. True pg. 93
2. True pg. 101
3. False pg. 101
4. True pg. 102
5. True pg. 104
6. True pg. 105
7. False pg. 104
8. False pg. 101
9. True pg. 99
10. False pg. 98

Fill-in

1. Alter, lesson objectives pg. 101
2. Instructor II pg. 101
3. Learning objective pg. 93
4. Level of instruction pg. 94
5. Adapt, modify pg. 101

Short Answer

1. Audience, behavior, condition, degree
2. Lesson title, level of instruction, behavioral objectives, instructional materials, lesson outline, references, lesson summary, assignment
3. Preparation, presentation, application, evaluation

Instructor Applications

Individual answers will depend on department operations and individual beliefs and impressions.

Chapter 7

Multiple Choice

1. A   pg. 117
2. B   pg. 117
3. C   pg. 118
4. D   pg. 121
5. A   pg. 121
6. D   pg. 122
7. C   pg. 123
8. C   pg. 124
9. D   pg. 118
10. A   pg. 121
11. B   pg. 122
12. D   pg. 122
13. A   pg. 122
14. C   pg. 123
15. B   pg. 125

Matching

1. G   pg. 116
2. C   pg. 117
3. B   pg. 119
4. H   pg. 119
5. D   pg. 121
6. J   pg. 121
7. E   pg. 122
8. A   pg. 123
9. I   pg. 123
10. F   pg. 124

True/False

1. False   pg. 119
2. True   pg. 119
3. True   pg. 117
4. False   pg. 117
5. False   pg. 118
6. True   pg. 118
7. False   pg. 121
8. False   pg. 122
9. False   pg. 123
10. True   pg. 125

Fill-in

1. culture   pg. 125
2. formal, behavior   pg. 125
3. contingency   pg. 124
4. personnel, safety   pg. 123
5. grin, in   pg. 117

Short Answer

1. Demographics includes race, national origin, age, gender, marital status, family size, and educational background. It could also include type of fire agency, career, volunteer, or paid on call (pg. 117).
2. Learning environment consists of the classroom, lesson plans, lesson outlines, and an evaluation tool (pg. 121).
3. A student should be comfortable in regards to classroom temperature, number of breaks during instruction, safety, creating a learning environment that encourages teamwork but allows for mistakes, and an environment where a student can be successful (pg. 122).

Instructor Applications

Answers to these questions will be based on personal experiences and local resources.

Chapter 8

Multiple Choice

1. A   pg. 130
2. B   pg. 131
3. C   pg. 131
4. D   pg. 132
5. B   pg. 132
6. A   pg. 132
7. C   pg. 135
8. D   pg. 135
9. A   pg. 136
10. B   pg. 138
11. C   pg. 139
12. D   pg. 142
13. A   pg. 143
14. B   pg. 144
15. C   pg. 144

Matching

1. G   pg. 130
2. E   pg. 131
3. I   pg. 132
4. B   pg. 134
5. C   pg. 135
6. F   pg. 136
7. A   pg. 137
8. H   pg. 137
9. D   pg. 139
10. J   pg. 140

True/False

1. True   pg. 140
2. False   pg. 140
3. True   pg. 140
4. False   pg. 142
5. False   pg. 142
6. True   pg. 143
7. True   pg. 143
8. False   pg. 144
9. True   pg. 145
10. True   pg. 145

Fill-in

1. Variety   pg. 131
2. Overreliance   pg. 132
3. Scanner   pg. 136
4. Digital audio player   pg. 138
5. Simulator   pg. 139

Short Answer

1. Understand the equipment that you will use during a presentation, and practice using it before your classroom presentation (pg. 131).
2. Check the connections between the computer and projector. Check the projectors light bulb, and make sure that the cover is off the lens of the project (pg. 145).
3. Multimedia appeals to more senses, which increases the retention of learning and keeps the learner involved in the learning process (pg. 144).

Instructor Applications

Answers to these questions will be based on personal experiences and local resources.

Chapter 9

Multiple Choice

1. A   pg. 150
2. B   pg. 150
3. C   pg. 150
4. D   pg. 151
5. A   pg. 152
6. B   pg. 152
7. C   pg. 154
8. D   pg. 154
9. A   pg. 157
10. B   pg. 157
11. C   pg. 158
12. B   pg. 158
13. C   pg. 159
14. C   pg. 159
15. D   pg. 159

Matching

1. F   pg. 150
2. I   pg. 151
3. G   pg. 151
4. C   pg. 152
5. D   pg. 152
6. E   pg. 153
7. J   pg. 155
8. B   pg. 157
9. A   pg. 153
10. H   pg. 154

True/False

1. True   pg. 150
2. False   pg. 150
3. True   pg. 152
4. True   pg. 152
5. False   pg. 154
6. True   pg. 154
7. False   pg. 155
8. False   pg. 155
9. True   pg. 157
10. True   pg. 157

Fill-in

1. empowered   pg. 151
2. technology, safety   pg. 151
3. creative   pg. 155
4. personnel   pg. 157
5. experience, drill ground   pg. 158

Short Answer

1. Departments should use the Life-Safety Initiatives in their polices and procedures to change the culture of the fire department to make it safer (pg. 150–151).
2. You teach safety as an instructor by example. You follow department policy in all of your training activities. You do not allow unsafe activities from those you train or those who assist you in the training process (pg. 152–154).
3. Safety is an attitude that becomes a part of you and how you do business. A good instructor learns to anticipate problems before they happen, and this comes from fire-ground experience, training, and good preparation (pg. 158–159).

Instructor Applications

Answers to these questions will be based on personal experiences and local resources.

Chapter 10

Multiple Choice

1. A   pg. 166
2. B   pg. 167
3. C   pg. 167
4. D   pg. 167
5. A   pg. 168
6. A   pg. 170
7. B   pg. 173
8. C   pg. 173
9. D   pg. 174
10. D   pg. 177
11. A   pg. 179
12. B   pg. 183
13. C   pg. 183
14. D   pg. 183
15. C   pg. 185

Matching

1. F   pg. 166
2. D   pg. 167
3. I   pg. 167
4. C   pg. 168
5. J   pg. 169
6. A   pg. 169
7. B   pg. 170
8. H   pg. 173
9. G   pg. 176
10. E   pg. 179

True/False

1. False   pg. 179
2. False   pg. 180
3. True   pg. 182
4. False   pg. 183
5. True   pg. 185
6. False   pg. 186
7. True   pg. 187
8. True   pg. 187
9. True   pg. 168
10. True   pg. 168

Fill-in

1. testing   pg. 166
2. written test   pg. 166
3. test analysis   pg. 168
4. arrangement   pg. 174
5. policy   pg. 185

Short Answer

1. Lack of standardized test specifications and test format. Confusing procedures and guidelines for test development, review, and approval. Lack of consistency and standardization in the application of testing technology. Failure to perform formal test-item analysis. Inadequate fire service instructor training in testing methods (pg. 166).
2. Face validity, technical content validity, job content validity or criterion reference validity, and currency of the information (pg. 168).
3. To evaluate the results to see that course objectives were met and that the questions measured the correct information (pg. 173).

Instructor Applications

Answers to these questions will be based on personal experiences and local resources.

Chapter 11

Multiple Choice

1.	A	pg. 192		9.	B	pg. 196
2.	D	pg. 192		10.	C	pg. 197
3.	C	pg. 193		11.	A	pg. 197
4.	B	pg. 193		12.	B	pg. 198
5.	B	pg. 194		13.	C	pg. 198
6.	A	pg. 194		14.	D	pg. 201
7.	C	pg. 194		15.	C	pg. 202
8.	D	pg. 196				

Matching

1.	F	pg. 192		6.	C	pg. 195
2.	E	pg. 192		7.	J	pg. 195
3.	D	pg. 192		8.	B	pg. 196
4.	I	pg. 194		9.	G	pg. 197
5.	H	pg. 194		10.	A	pg. 196

True/False

1.	False	pg. 193		6.	True	pg. 197
2.	True	pg. 198		7.	True	pg. 196
3.	True	pg. 192		8.	False	pg. 196
4.	True	pg. 193		9.	False	pg. 196
5.	False	pg. 193		10.	True	pg. 198

Fill-in

1. road map — pg. 196
2. annual, complacent — pg. 196
3. skill set — pg. 197
4. safety — pg. 198
5. professional — pg. 201

Short Answer

1. Policy should contain the evaluation tool or how to obtain one or develop it. The policy should also address when evaluations should be given, by whom, for what purpose and then what to do once the evaluation is complete, including feedback to the instructor (pg. 197–201).
2. Before evaluating an instructor, you should review the lesson plan and content, obtain the evaluation tool, and review its content. You should review department policy for conducting an evaluation. You should determine the purpose of the evaluation and follow department policy regarding the reason for the evaluation. You should arrive early, introduce yourself to the instructor, and let him or her know the reason you are in his or her class. Complete the evaluation form during the allotted time. After the form is completed, follow your department's policy for providing feedback to the instructor (pg. 197).
3. After you have evaluated an instructor, you should provide feedback to them. In doing so, you should key in on the positives first and then, as appropriate, identify areas that need to be improved. When offering suggestions, do so in a friendly yet firm manner and offer any ideas on how to improve and be supportive. This is an opportunity to coach an instructor to make him or her better and increase his or her skills in teaching the next generation of fire fighters (pg. 198–201).

Instructor Applications

Answers to these questions will be based on personal experiences and local resources.

Chapter 12

Multiple Choice

1. A   pg. 208
2. B   pg. 208
3. C   pg. 208
4. D   pg. 209
5. A   pg. 209, 210
6. A   pg. 213
7. B   pg. 213
8. C   pg. 214, 215
9. C   pg. 216
10. D   pg. 217
11. A   pg. 221
12. B   pg. 225
13. C   pg. 226
14. D   pg. 213
15. A   pg. 209

Matching

1. I   pg. 209
2. E   pg. 210
3. C   pg. 210
4. H   pg. 210
5. J   pg. 213
6. F   pg. 217
7. A   pg. 217
8. D   pg. 217
9. G   pg. 218
10. B   pg. 221

True/False

1. True   pg. 224
2. True   pg. 223
3. False   pg. 213
4. False   pg. 208
5. False   pg. 214
6. False   pg. 209
7. True   pg. 210
8. False   pg. 211
9. True   pg. 213
10. True   pg. 217

Fill-in

1. safety   pg. 208
2. management, schedule   pg. 208
3. attended, participated   pg. 213
4. respiratory protection, infectious disease protection   pg. 214
5. budgeting   pg. 217

Short Answer

1. Determine what is to be taught, who your audience is, and where and when the training should take place. Arrange for a qualified instructor, and have the training approved if necessary. Send out notice to those who will be required to attend, and ensure that all necessary resources will be available (pg. 216).
2. Review previous budgets, identify needs and required resources, prepare budget given revenues and expenditures of your department, and submit your budget for internal and external review. Once the budget is approved, follow budget and monitor its use. At the appropriate time, close out current budget and begin preparations for next year's budget (pg. 217–219).
3. Select an instructor that is qualified as an instructor and has a knowledge base of the subject to be taught. Once the instructor is approved, supply him or her with lesson material and resources for the course to be taught (pg. 223–224).

Instructor Applications

Answers to these questions will be based on personal experiences and local resources.

Chapter 13

Multiple Choice

1.	C	pg. 232		9.	C	pg. 236
2.	B	pg. 232		10.	A	pg. 236
3.	A	pg. 233		11.	D	pg. 238
4.	D	pg. 233		12.	B	pg. 238
5.	B	pg. 233		13.	C	pg. 239
6.	C	pg. 233		14.	B	pg. 239
7.	B	pg. 234		15.	D	pg. 239
8.	A	pg. 234				

Matching

1.	F	pg. 239		6.	B	pg. 236
2.	H	pg. 239		7.	A	pg. 234
3.	C	pg. 239		8.	D	pg. 234
4.	C	pg. 236		9.	J	pg. 236
5.	I	pg. 236		10.	G	pg. 233

True/False

1.	True	pg. 232		6.	False	pg. 234
2.	True	pg. 233		7.	False	pg. 235
3.	False	pg. 233		8.	False	pg. 236
4.	True	pg. 233		9.	False	pg. 239
5.	False	pg. 234		10.	True	pg. 239

Fill-in

1. well-rounded — pg. 233
2. promotion, bachelors — pg. 234
3. Networking — pg. 234
4. teaching skills — pg. 236
5. peplaced — pg. 236

Short Answer

1. Lifelong learning is the basic idea that you should always be looking for opportunities to learn and grow. This can be done by networking, reading articles, or publishing books. As long as you are learning something new, you will continue to grow (pg. 233).
2. By mastering the use of your time through setting goals and planning your daily activities, you learn to control your time. Good time management skills require that you set priorities and discipline yourself to follow them but remain flexible to adapt to schedule changes if necessary (pg. 233).
3. Look for individuals within your organization who could be developed into the next generation of instructors and begin the development process with them (pg. 236).

Instructor Applications

Answers to these questions will be based on personal experiences and local resources.

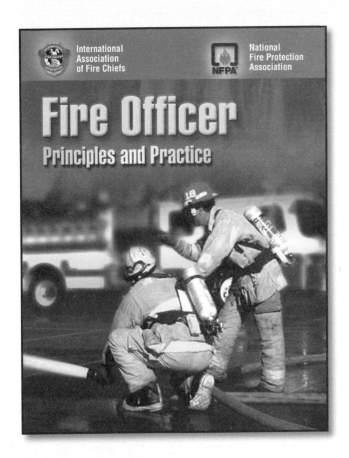

Fire Officer
Principles and Practice

International Association of Fire Chiefs
National Fire Protection Association
ISBN-13: 978-0-7637-2247-0
Paperback • 414 Pages • © 2006

Fire Officer: Principles and Practice is the core of the teaching and learning system, with features that reinforce and expand on the essential information and make information retrieval a snap. Covering the entire scope of the current edition of NFPA 1021, *Standard for Fire Officer Professional Qualifications*, ***Fire Officer: Principles and Practice*** covers everything from fire officer communications to managing fire incidents.

Other topics include:

- Preparing for promotion
- Training and coaching
- Working in the community
- Creating a budget
- Organized labor and the fire officer
- Crew resource management
- Communications and presentation skills

Order ***Fire Officer: Principles and Practice*** today at 1-800-832-0034 or www.jbpub.com/Fire.

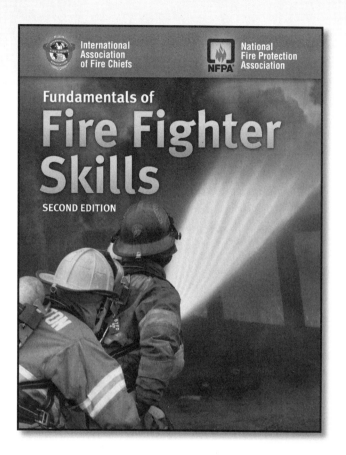

Fundamentals of Fire Fighter Skills
SECOND EDITION

International Association of Fire Chiefs
National Fire Protection Association
ISBN-13: 978-0-7637-7145-4
Paperback • 1068 Pages • © 2009

With the release of the *Second Edition*, Jones and Bartlett Publishers, the National Fire Protection Association®, and the International Association of Fire Chiefs have joined forces to raise the bar for the fire service once again.

The *Second Edition* features a laser-like focus on fire fighter injury prevention, including a dedicated chapter on safety. Reducing fire fighter injuries and deaths requires the dedicated efforts of every fire fighter, of every fire department, and of the entire fire community working together. It is with this goal in mind that we have integrated the 16 Fire Fighter Life Safety Initiatives developed by the National Fallen Fire Fighter Foundation into Chapter 2, Fire Fighter Safety. In most of the chapters, actual National Fire Fighter Near-Miss Reporting System cases are discussed to drive home important points about safety and the lessons learned from those real-life incidents. It is our profound hope that this textbook will contribute to the goal of reducing line-of-duty deaths by 25 percent in the next 5 years.

Fundamentals of Fire Fighter Skills, Second Edition thoroughly supports instructors and prepares students for the job. This one-volume text meets and exceeds the Fire Fighter I and II professional qualifications levels as outlined in the 2008 edition of NFPA 1001, *Standard for Fire Fighter Professional Qualifications*. It also covers all of the Job Performance Requirements (JPRs) listed in the 2008 edition of NFPA 472, *Standard for Competence of Responders to Hazardous Materials/Weapons of Mass Destruction Incidents*, at the awareness and operations levels, including Section 6.2, Mission-Specific Competencies: Personal Protective Equipment and Section 6.6, Mission-Specific Competencies: Product Control.

Order ***Fundamentals of Fire Fighter Skills, Second Edition*** today at 1-800-832-0034 or *www.jbpub.com/Fire*.

www.jbpub.com/Fire

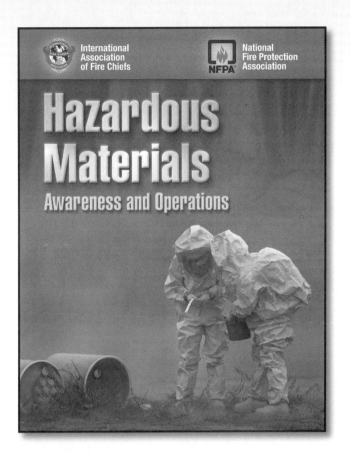

A fire fighter's ability to recognize an incident involving hazardous materials is critical. They must possess the knowledge required to identify the presence of hazardous materials and weapons of mass destruction (WMD), and have an understanding of what their role is within the response plan. *Hazardous Materials: Awareness and Operations* will provide fire fighters and first responders with these skills and enable them to keep themselves and others safe while mitigating these potentially deadly incidents.

Hazardous Materials: Awareness and Operations is the center of an integrated teaching and learning system that combines groundbreaking content with dynamic new features to support instructors and to help prepare students for the job. The text meets and exceeds the requirements for Fire Fighter I and II certification and satisfies the core competencies for operations level responders including the eight mission-specific responsibilities for first responders within the 2008 Edition of NFPA 472, *Standard for Competence of Responders to Hazardous Materials/Weapons of Mass Destruction Incidents*. Additionally, the material presented also exceeds the hazardous materials response requirements of the Occupational Safety and Health Administration (OSHA) and the Environmental Protection Agency (EPA).

Hazardous Materials: Awareness and Operations provides in-depth coverage of:

- The properties and effects of hazardous materials and WMDs
- How to calculate potential danger and initiate a response plan
- Selection, use, advantages, and disadvantages of personal protective equipment.
- Performing mass and technical decontamination.
- Performing evidence preservation and sampling.
- Performing product control.
- Performing air monitoring and sampling.
- Performing victim rescue and recovery.
- Responding to illicit laboratory incidents.

Hazardous Materials
Awareness and Operations

International Association of Fire Chiefs
National Fire Protection Association
Robert Schnepp
ISBN-13: 978-0-7637-3872-3
Paperback • © 2010

Order **Hazardous Materials: Awareness and Operations** today at 1-800-832-0034 or *www.jbpub.com/Fire*.